Praise for *Raising Rabbits for Meat*

As our culture moves toward more honest sources of meat, the rabbit continues to be one of the most efficient and nutritious options. Eric and Callene Rapp have a wealth of experience which they share in this detailed and personable guide. The Rapp's dedication to breed integrity and to rabbit health are exceptional, and this book will be essential for anyone interested in raising rabbits—from the homesteader to the commercial farmer.

— Meredith Leigh, author, *The Ethical Meat Handbook* and *Pure Charcuterie*

So you think you want to raise rabbits?!? Everything you need to know to do it successfully and enjoyably is contained in *Raising Rabbits for Meat*, co-authored by the most comprehensive rabbit-rearing couple of our time, Eric and Callene Rapp. Readers will not only receive information gleaned from decades of experience but they will become engrossed in the Rapps' engaging delivery. *Raising Rabbits* is destined to become the single authority on all aspects of creating and maintaining a successful meat rabbit business and should be on the bookshelf of every livestock farmer out there and those who aspire to raise rabbits, poultry, or other stock as a business.

— Hank Will, Editorial Director, Ogden Publications

You may have noticed that I've never written about raising rabbits. That's because it was a complete disaster when we tried—with three different breeds! If only I'd had this book back then, we might have been enjoying homegrown rabbit all these years. The Rapps tell you everything you need to know, from setting up your rabbitry to buying your first rabbits, and from breeding and kindling to harvesting.

— Deborah Niemann, ThriftyHomesteader.com, author, *Homegrown and Handmade*, *Ecothrifty*, and *Raising Goats Naturally*

Kudos for *Raising Rabbits for Meat*. It is a useful tool that is fully loaded with the knowledge that newcomers will need for success in building a rabbitry from the ground up. The common-sense practical advice coupled with the Rapp's notorious wit make the book a delightful read.

— Jeannette Beranger, Senior Program Manager, The Livestock Conservancy

RAISING RABBITS
FOR MEAT

Eric Rapp & Callene Rapp

new society
PUBLISHERS

Cover design by Diane McIntosh.
Cover images: top and bottom rabbit photos: Callene Rapp; background texture:
© iStock 517263067; rabbit illustration: © iStock 494058002.
All photos figures © Eric and Callene Rapp unless otherwise noted.
Interior images: p. 1, 5, 13, 19, 31, 41, 61, 79, 127, 141, 149 © doublebubble_rus 5282;
p. 87 © jenesesimre; p. 95, 119 © unorobus; p. 107 © morningglory1285 / Adobe Stock.

Printed in Canada. First printing September 2018.

This book is intended to be educational and informative.
It is not intended to serve as a guide. The author and publisher disclaim all
responsibility for any liability, loss or risk that may be associated with the application
of any of the contents of this book.

Inquiries regarding requests to reprint all or part of *Raising Rabbits for Meat*
should be addressed to New Society Publishers at the address below.
To order directly from the publishers, please call toll-free
(North America) 1-800-567-6772, or order online at www.newsociety.com

Any other inquiries can be directed by mail to:

New Society Publishers
P.O. Box 189, Gabriola Island, BC V0R 1X0, Canada
(250) 247-9737

LIBRARY AND ARCHIVES CANADA CATALOGUING IN PUBLICATION

Rapp, Eric, 1959–, author
Raising rabbits for meat / Eric Rapp & Callene Rapp.

Issued in print and electronic formats.
ISBN 978-0-86571-889-0 (softcover).—ISBN 978-1-55092-682-8 (PDF).—
ISBN 978-1-77142-278-9 (EPUB)

1. Rabbits.   2. Meat animals.   I. Rapp, Callene, 1965–, author
II. Title.

SF453.R37 2018   636.932'2   C2018-904554-X
                                                       C2018-904555-8

Funded by the   Financé par le
Government   gouvernement
of Canada   du Canada

New Society Publishers' mission is to publish books that contribute in fundamental
ways to building an ecologically sustainable and just society, and to do so with the
least possible impact on the environment, in a manner that models this vision.

This book is dedicated to all the rabbits
who have nourished us, challenged us, and taught us,
and all the master breeders who passed down their knowledge.
We are all better for having known you.

Contents

Acknowledgments

While it's true that the act of writing is a solitary endeavor, no book ever sees the light of day without a host of other people involved in the effort.

Eric is the one who, day in and day out for the last 15 years, has cared for our rabbits with meticulous attention to detail. I am merely the scribe who attempted to put all that experience down on paper.

To the chefs who have been our partners throughout this little adventure—we can't thank you enough.

Thanks to my friend, Danielle, who helped transcribe citations when I was too bleary-eyed to focus, and thanks to everyone who read any part of this manuscript to answer the question, "Does this make sense?"

Thank you to Jordan for her beautiful artwork, and thanks to the Graphics staff at SCZ for helping me scan them into a format that would actually work in the book.

Thanks to Ingrid from New Society Publishers for believing in this book, and for dealing with the shenanigans and the cluelessness of a first-time author.

And special thanks to New Society Publishers copy editor Betsy Nuse. The fact that you are actually holding a physical copy of a book in your hands is in no small part due to her efforts in shepherding, guiding and helping organize what at many points in the process could only be described as a "hot mess." She assured me it was not, but I think that was just that innate Canadian politeness talking.

—Callene

CHAPTER 1

The History of Rabbit Keeping

Among the major species of domesticated animals, rabbits were rather late to the party.

The fossil record of rabbits and their ancestors is a bit of an anomaly. There are more distinct species in the fossil record than exist in the living world today. New techniques in recovering fossils have no doubt contributed to scientist's ability to recover the delicate bones and teeth that identify the ancestors of our modern rabbits.[1]

But it is without a doubt that one of the most widespread species of rabbit in the world is the European wild rabbit, also known as *Oryctolagus cuniculus*, the ancestor of our modern domestic rabbit.

In Europe, rabbits were first described by Phoenician sailors about 1000 BCE. The sailors were amazed by the tiny animal's extensive burrowing ability, and they brought tales about them back to their home ports. They called the land where they had discovered rabbits *I-Saphan-Im*, which translated into High Latin as *Hispania*, and later became the Spanish word *España*.

Thus, the very name of Spain is linked to rabbits. Spanish coins in Roman times even featured rabbits on them. Romans then seem to have spread rabbits extensively throughout their Empire, mostly as a game animal.[2]

The first writings mentioning rabbits as something other than wild animals are found in the work of the Roman historian Marcus Terentius Varro (116–27 BCE). Varro advocated putting rabbits in walled *leporaria* (rabbit gardens) to facilitate hunting. These weren't gardens

as we commonly think of them, but large parks ranging up to a hundred acres or so.[3]

And while these *leporaria* are the origin of the warren system of rabbit keeping, the rabbit was still not a truly domesticated animal, although they most likely had become tamed (desensitized to human presence). A figurine from the third century CE found in France depicts a child holding a young rabbit; we can presume, from evidence like this, that interactions between humans and rabbits were becoming more common.[4]

In southern France, archaeologists have discovered *cadaver wells* (what seem to be the ancient equivalents of trash pits) near clusters of homes. The skeletal remains of rabbits are present in these pits in high enough concentration to indicate that rabbit was a readily available and presumably common menu item.[5]

Romans adopted the ancient Spanish custom of eating *laurices* (fetal or newborn rabbits) helping to spread that custom throughout the empire so widely that laurices eventually became a well-known food during the Lenten period in France.[6]

Eventually, keeping rabbit warrens in France became the sole right of the nobility. Professional warreners were employed to manage the animals, both in the warren itself and to prevent the animals from escaping and damaging neighboring fields and crops.[7]

Monasteries in Western Europe began keeping rabbits during the medieval period, and records from that time exist of rabbit trading between the monks and nobles.[8]

By the 13th century, the only mentions of rabbits were of those kept in warrens as we think of them, and where intentional selection for specific traits takes place. It would not be much of a stretch to picture monks noting different colors or patterns that might pop up from time to time, pulling those animals from the warren and bringing them into a more tightly managed breeding system to concentrate those differences.[9]

Very little writing exists on any of those efforts or management details, but by the 16th century, several color varieties of rabbit are described in written records. The Champagne d'Argent, one of the oldest breeds of rabbit in the world, was developed by monks in the Champagne region of France. The name literally means "Silver from Champagne."[10]

By the 18th and 19th century, rabbit keeping was no longer the sole privilege of the nobility, and more writings on rabbit husbandry begin to emerge. Rabbit hutches sprang up all over Western Europe, both in rural settings and in towns. Rabbits were kept as a ready source of meat which could be accessed as needed. These animals were fed on forage picked daily, as well as crops such as grains, roots, and hay.[11]

Interestingly, there seems to have been a reduced amount of rabbits produced after that change, possibly due to the change in, or inconsistency of, diet provided.

In the late 1890s, Belgian Hares (which are really a rabbit, not a hare—see Chapter 3) were imported to the United States, and the rabbit craze was on. Rabbits sold for unheard-of prices at the time, and families such as Guggenheim and Rockefeller were well-known figures in the American rabbit world.

By the early 20th century, a veritable explosion of breeds and varieties occurred. Creating new breeds and strains became a serious practice for fanciers at the time, and several of the breeds we know today (such as the American Blue and White, the American Chinchilla, and the Silver Fox) were created.[12]

Rabbits became big business. In fact, Edward Stahl, the founder of the American Chinchilla breed, still bears the distinction of being the only person to make a million dollars with rabbits, during the Great Depression. An ad in

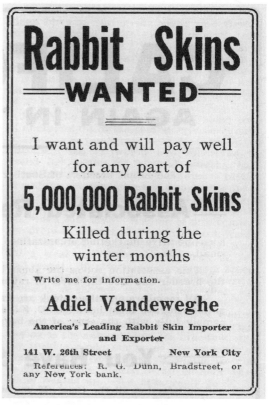

FIGURE 1.1. This ad is one of several in that particular issue. We always find old magazines like this fascinating for the look they offer at management and techniques of days gone by.

a *Hares and Rabbits* magazine of the time advertised *one* furrier looking to purchase five million rabbit pelts.[13] Imagine if there was a market for that many pelts, how many rabbits existed at the time!

In 1928, the first research station in the United States dedicated specifically to rabbits was built in Fontana, California, and produced a lot of the information about rabbit husbandry we rely on today. Unfortunately, this research facility was closed in 1964 due to a lack of funding.[14]

The work of the research station led to new methods of rabbit raising, most specifically the all cage broiler system (which all but eliminated the problem of *coccidia*) and the use of pelleted feeds (which eliminated a lot of the labor of raising rabbits and ensured that each rabbit got a balanced diet). Several breeds were also created and selected for production traits.

Changes in the food system after World War II favored more intensive, large-scale meat production with animals such as cattle and pigs, and rabbit fell out of favor. Fortunately, people are rediscovering how useful rabbits are on the homestead, which is what may have brought you to this book.

Rabbits can be raised in a variety of environments, can adapt to nearly any management scheme, and can provide a plethora of useful products. Small and quiet, they require much less in the way of fencing and space, but give back well beyond what they require from us.

Modern rabbit keeping differs greatly from the ancient *leporaria*, but rabbit still remains a unique and very useful livestock.

CHAPTER 2

What's Your Plan?

This is a question Eric asks me regularly when I hatch another idea, or seem to be jumping into any scheme too quickly or without a lot of forethought. It's a joke around our house, but "what's your plan?" is a valid question, and one that should be asked and re-asked, on a regular basis.

What is your ultimate goal when beginning your rabbit venture? Is it *self-sufficiency*? No other species of livestock will produce quality protein with less space than the rabbit, and rabbits will add additional benefit to your garden by producing the best fertilizer available. Is it a niche at the *farmers' market*? Rabbit, for some folks, can take a little getting used to, but once market shoppers have tried it, they will be loyal customers for life. Is it marketing to *restaurants*? A new generation of bold and creative chefs have made rabbit a hot menu item across the US. Are you interested in *heritage breed conservation*? All heritage breeds need good, solid *breed stewards* willing to put their own egos aside and work to the benefit of the breed as a whole.

There is no right or wrong goal. But be honest with yourself about your skills, abilities, and resources—especially time.

It's easy to get caught up in the excitement of starting a rabbitry, but without a good solid goal it can be too easy to wind up with a mess. Don't be like one well-meaning family that purchased rabbits with the goal of supplying meat for the family...and found out, when it came time, they could not process the rabbits. Now these folks have a dozen, beautiful, well-cared-for pets.

The assumption of *Raising Rabbits for Meat* is that you are interested, on some level, on producing and processing meat rabbits. There

FIGURE 2.1. This USDA label provides quality assurance.

is no way from point A to point B without harvesting those animals. The best time to have the conversation about whether or not you are able to kill and process animals yourself is well before you make the first mating.

On the flip side, if your goal is to produce fryers for sale either to restaurants or farmers' markets, do some research into both federal and state regulations.[1] Whether or not you like it, the law is the law, and trying to circumvent legalities one way or another is a risky, potentially lose-it-all strategy. In the US, USDA inspected processing is required for sale to restaurants and for shipping across state lines, and state regulations can vary widely on how many rabbits can be sold at farmers' markets.

Self-Sufficiency

Raising rabbits for your own freezer is a highly fulfilling job. You can feel satisfaction like no other to be able to look in a freezer full of home processed meat and know that you are doing the best you can to take control over your food supply.

If you plan to involve your children in raising rabbits, have a good long conversation with them about your goals, plans, and what you expect their involvement to be. Many parents dread having an honest conversation with their kids about what the ultimate destination of those rabbits will be, but often that dread can lead to making too big deal out of it and creating the problem you fear solving.

After all, barely a generation ago, all our food came from either the farm or a local source, and kids helped feed, water, and care for animals every step of the way. We haven't evolved that far from these roots, we've just lost touch with them. I firmly believe that we aren't doing kids a favor by sugarcoating the realities of life and death, including what it takes to eat meat for dinner. I also believe that kids are brighter and more resilient than we often give them credit for.

Processing at home is probably the largest hurdle for the novice. See if you can find someone experienced to walk you through it the first couple of times, and don't be shy about asking for help.

Farmers' Markets

If one of your rabbitry goals is to make a little money, farmers' markets may be a good option. Depending on rules and regulations where you live, government inspection may be required to sell retail, especially if you plan to sell over a certain quantity. Be sure to include costs such as processing, transportation to and from market, booth rental, and other miscellaneous costs in pricing your meat per pound. Also, factor your time into the cost as well. Often truly pricing out what it costs to produce that fryer can yield a price per pound that may shock consumers who are used to cheap commodity meat. Yes, today's consumers are much more educated, but there is still a way to go.

Another wise investment if you decide to sell rabbits for meat is product liability insurance. Anyone can be named in a lawsuit. No one likes paying for insurance. But in today's increasingly litigious society, insurance can be an inexpensive cost for peace of mind.

A piece of advice: Don't name every rabbit in every litter. Name the breeding stock if you must, but give yourself a little distance, at least initially.

Restaurants

Seeing a rabbit dish on a menu for $20–$30 per plate can give you dollar signs for eyeballs. Keep in mind that government-supervised processing of meat served to the public is required, which can add a lot to the costs of production. Restaurants also depend on being able to acquire a consistent volume of rabbit. If they plan to put you on the menu and one month you can provide 25, but the next month only 12, they will find another producer, and they may or may not come back to you.

Rabbit, just because of its size to cost ratio, is one of the more expensive proteins chefs will ever use. Their bottom line is also critical for both your success. So don't shortchange yourself to get your rabbit on

FIGURE 2.2. A rabbit dish at the Rieger Restaurant in Kansas City, Missouri.

their menu; a good product will speak for itself in the hands of a good chef.

Many chefs have also become accustomed to being able to pick up the phone and order what they need at a moment's notice. For some it can be a rude awakening to discover that you may or may not have a ready supply of fryers.

And likewise, chefs can change their menu without notifying you. If you've raised to a certain production level based on their order and they stop ordering, you have a surplus you now need to scramble to market.

A talented chef that understands production cycles, variability, open and honest communication, and is invested in your product is a valuable asset. Treat those chefs accordingly.

A wise, longtime breeder once told us that in order to be profitable marketing heritage meat products, you either have to stay so small your costs stay small as well so that you can control the whole process— or you must become so large that you drive the market. Restaurant sales are a middle ground that can eat into profit quite quickly, and you can find yourself being forced to be more reactive than proactive.

We wouldn't change a thing about our journey because it has led us to friendships within the food community that we truly value, but there's no doubt a different path would have been easier.

Heritage Breed Conservation

Our focus at Rare Hare Barn has been conservation of heritage breeds. The meat business came about as a result of the conservation mission, rather than the other way round.

The breeds we chose to focus on at one time constituted a large part of the millions of rabbits found in the US in the early 20th century. These rabbits were dual-purpose breeds, selected for both meat and fur quality, and the animals are larger than many of the breeds in vogue more recently.

With the rise of industrial agriculture and the advent of subsidized meat production—and with an increasingly urban population—rabbit declined in popularity and availability as a homestead meat animal. As a fur animal, the focus turned to rabbits with white pelts which could easily be dyed, and white rabbits such as the New Zealand began to replace colorful fur breeds such as the Silver Fox or the American Blue.

Coupled with the explosion of pet breeds and the trend to consider rabbits more as pets than livestock, certain breeds began to decline precipitously in number. By the late 1990s, some breeds which had once numbered in the hundreds of thousands had been reduced to only a few hundred.

At one American Rabbit Breeders Association (ARBA) gathering in Wichita, Kansas, among the nearly 20,000 rabbits and cavies (guinea pigs) on exhibit, I struggled to find five American Blue and White rabbits, stuffed in a corner of the exhibition hall.

This may on the surface not seem like a big deal, but recognition on the ARBA list of breeds is determined by each breed's presence at a national show. Should a breed fail to exhibit in five consecutive shows, it will be dropped from the list of recognized breeds. Once dropped, an entirely new Certificate of Development must be issued, and it can take years, and multiple tries, to get the breed accepted again. This is such a laborious process that once a breed or variety has been dropped, it is likely to not be added back, and may roll down the cliff to extinction.

Finding only that handful of rabbits cemented our desire to work with scarce breeds, even though our focus was not showing. These breeds were outstanding production rabbits back in the day, and we felt that they could be again.

Heritage breeds are also more likely to require a little more outlay in cash to purchase, and some travel may be required to find them. Make sure the selling breeder is able to provide pedigrees for stock as well.

Registering a rabbit is also different from registering offspring from other species. To be registered, an ARBA certified registrar will inspect the rabbit, to make sure both it and its ancestors reached the proper weight, and then if the rabbit passes, the registrar will place a tattoo in the rabbit's right ear.

This is not important for meat production, but should you decide to get in to rabbit showing, it may be something you wish to consider.

Breed Stewardship

The phrase *breed steward* gets tossed around quite a bit these days. Heritage breeds are often exciting to those who want to raise something different, unique, and challenging.

But unfortunately, that uniqueness can also become a breeder's downfall rather quickly. Heritage breeds, rabbit and otherwise, don't fit standard marketing schemes very well. This can be somewhat daunting to those who believe that "if you build it, they will come."

Our informal survey of people getting into heritage breeds (no matter what species) found that the average length of time some new breeders stick with the animals is about 18 months.[2] This gets them through one round of seasons, a couple of production cycles, and into the phase where the invariable challenges present themselves, and their management has the most impact on production.

Most of the time in these cases, I believe people have truly just not done their homework regarding what is involved in committing to the breed or species over a long term. And heritage breed stewardship is a commitment.

A true breed steward understands:
- That stewardship is a lifelong commitment.
- That the highs and lows level out over that lifetime.
- It's not a get-rich-quick scheme.
- It means making tough decisions and placing what benefits the breed as a whole over personal desires and goals.

True commitment to the breed and the breed standard is necessary. Don't try to change the breed or breed standard to suit you. There is a huge difference between selecting traits that suit your production goals and trying to reinvent the breed altogether. Don't try to change the breed standard to suit the animals you are breeding; breed better animals true to the breed standard.

Should the breed or species not work out, put some time and effort into passing them along to another breeder rather than dumping them at a sale barn, especially in the case of a critically endangered breed.

Breed stewardship isn't for the fainthearted. But true stewards who are invested in the long-term success of the breed are crucial to its survival.

Helpful Pointers

As with all other species of livestock, management is the key to long-term success.

The best stock, the best equipment, and the best intentions won't make up for poor or inconsistent management. Rabbits may require less in the way of equipment and facilities than other livestock, but they require as much, if not more, skill and attention to detail.

While rabbits are easier for an individual (or for children) to take care of by themselves, good husbandry skills are essential. Rabbits can be low maintenance, but they are not no maintenance.

Observation is key. Good observation and consistent care and management will take your rabbit venture to the next level, and keep your rabbits healthy and productive.

CHAPTER 3

Rabbit Biology

Rabbits are members of the taxonomic order *Lagomorpha*, animals which have four incisor teeth in the upper jaw, and two in the lower. Up until the early 20th century, rabbits and hares were believed to be in the same order as rodents, but rodents only have two incisor teeth in the upper jaw, so the two groups were separated. Rabbits are also totally herbivorous, whereas rodents will eat meat.

Order *Lagomorpha* is further divided into the families *Leporidae* and *Ochotonidae*. *Leporidae* translates into "those that resemble hares" and is a large family, containing many genera and over 60 species.

The *Leporidae* family breaks down into genera and subgenera.

Three genera relevant to rabbit keeping are the genus *Lepus* (which includes hares and jackrabbits), the genus *Sylvilagus* (which includes cottontails and North American wild rabbits), and the genus *Oryctolagus*, which is the European wild rabbit (*Oryctolagus cuniculus*), the progenitor of all our modern breeds of domesticated rabbit.

No matter how much rabbits and hares resemble one another, they cannot interbreed. Domestic rabbits let loose will not run off to join their wild cottontail cousins and make happy little hybrid babies.

Each of the three genera have different numbers of chromosomes: Hares have 24 pairs, cottontails have 23 pairs and domestic rabbits have 22. No controlled lab studies have produced offspring from breeding rabbits and hares, and no matter how many anecdotal stories exist, none have been verifiable.[1]

> **Rabbit Taxonomy**
> ▸ **Kingdom:** *Animalia* (not plant or mineral)
> ▸ **Phylum:** *Chordata* (has a spinal cord)
> ▸ **Class:** *Mammalia* (warmblooded, give birth to live young, have hair)
> ▸ **Order:** *Lagomorpha* (having four incisor teeth in the upper jaw)
> ▸ **Family:** *Leporidae* (nearly 60 species of rabbit and hare)

Characteristics

Rabbits and hares differ in a few important physiological ways. It doesn't help clarify things that there is a breed of rabbit called the Belgian Hare and a species of hare called the Jackrabbit. However, the young of rabbits and hares are very different at birth, and they tend to have different athletic abilities and thrive in different environments.

Rabbit young are *altricial*—meaning born hairless, with their eyes closed, and completely dependent on their mother's ability to make a warm and cozy nest.

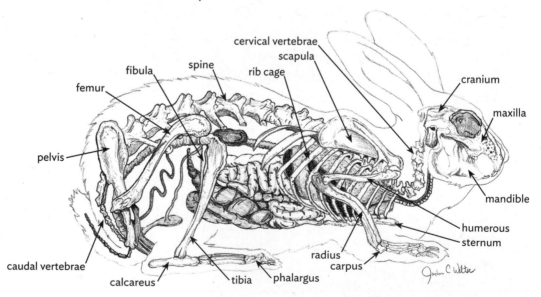

FIGURE 3.1. Rabbit skeleton. Credit: Jordan Wiltse.

Rabbits (*Oryctolagus* and *Sylvilagus*)
- born hairless
- born blind
- born helpless in a fur-lined nest
- live in brush and cover (wild)
- have shorter legs and are built for quick bursts of speed
- have relatively shorter ears

Hares (*Lepus*)
- born fully haired
- born with eyes open
- can run within a few minutes of birth
- born in a grassy depression in the ground
- have long legs and are built for endurance
- have very long ears

Hares on the other hand are *precocial*—born fully haired, with eyes open, and are able to get around soon after birth.

Hares are built for not only speed but endurance, and thrive in areas with sparse plant life. Rabbits bank on quick bursts of speed and the ability to outmaneuver predators, and tend to be found in areas with more dense vegetation.

For the most part, hares and rabbits look a great deal alike so it's easy to mistake hares for rabbits and vice versa. They both generally have *agouti coloration*, which is a pattern of banding on each individual hair shaft that gives them the familiar brown, peppered look.

Distribution

Rabbits and hares are found on nearly every continent. They traveled to most areas of the world, either by natural migration or by being intentionally released on islands and in other areas to serve as a future source of meat. On many islands, in the absence of any natural predators, rabbits thrived at the expense of the local flora and fauna, devastating natural ecosystems.

Australia is a poignant example. A handful of rabbits were released in the late 1850s for hunting purposes. Those few rabbits multiplied rapidly, and a decade later, millions were present, representing one of the fastest population explosions of any mammal in history.

Species versus Breed

All of our breeds and varieties of domestic rabbit descended from the European wild rabbit, *Oryctolagus cuniculus*, which also started out as a plain agouti-colored animal, and through careful (and sometimes accidental) selection, took on the variety of colors we see today. Breed is a further subclassification of species.

As noted earlier, different species of Lagomorphs cannot interbreed, but the different breeds of domestic rabbit *can* because they are the same species.

These rabbits have had a devastating effect on Australia's ecology, eating native plants that kept topsoil from eroding; they have likely inadvertently caused the extinction of numbers of native animals as well. Control efforts, which have included poison, trapping, hunting, and the introduction of the Myxoma virus, have all met with limited success.[2]

Fortunately in North America, the abundance of native predators helped keep rabbit populations in check.

Rabbit Physiology

If you have ever picked up and held a rabbit, you have more than likely felt its rapid heart rate.

A rabbit at rest can have a heart rate of 130 beats per minute, but an active rabbit, or one that is stressed, can have a rate of up to 325 beats per minute. Their body temperature ranges from 99.1 to 102.9°F (37.3 to 39.4°C), and respiration ranges from 30 to 60 breaths per minute. Rabbits also as a rule breathe through their noses, and to find a rabbit breathing through its mouth is generally a very bad sign.

As you might expect from an animal with a metabolism that fast, rabbits are not a long-lived animal. Pet rabbits can live five to eight years, some even longer. Most production rabbits will remain in the herd for three or four years, after which they will generally be culled because their production has dropped. It sounds harsh, but because an important goal of a business is to turn a profit, feeding unproductive animals makes poor economic sense. We have on occasion had production rabbits well outlast their contemporaries. One particular doe was productive for six years, and we currently have a buck that recently turned seven. Both of these rabbits passed their productivity on to their offspring, and have earned their pensions. But individuals such as this are the exception, rather than the rule.

Rabbits also have one of the fastest turn-arounds (meaning how fast they can go from birth to harvest) of any domestic animal. With a relatively short gestation and incredibly fast-growing offspring, a rabbit can become meat on the table in 12 weeks or less compared to a year or more for other species.[3]

Rabbits possess one of the lightest skeletal systems of any mammal, averaging around 7–8 percent of their total body weight (compared to 15 percent in a cat of comparable size). This,

> **Rabbit Physiology**
> ► **Lifespan:** (average) 5–6 years
> ► **Temp:** 99.1–102.9°F (37.3 to 39.4°C)
> ► **Pulse:** 130–325 beats per minute
> ► **Respiration:** 30–60 breaths per minute
> ► **Estrous:** continuous/seasonal
> ► **Gestation:** 30–33 days
> ► **Wean:** 4–6 weeks

combined with the incredibly powerful muscles in the hindquarters, is the main reason a rabbit can injure itself, even breaking its own back, if not handled properly. A misplaced kick, or a twist when being picked up, is enough to do the trick. This is why it's so important to learn to properly handle rabbits and be able to confidently and securely pick them up and carry them. It's pretty devastating to have a promising animal be ruined when injury could have been prevented. The occasional freak accident can happen, though: once Eric placed a doe in a buck's cage for mating. She stood up on the wall of the pen, just as the buck was attempting to mount, and the force of his body combined with the odd angle of hers was enough to break her back. However, this freak accident has only happened once in thousands and thousands of matings. Don't be afraid of handling your rabbits, just be aware of the vulnerability of that back. Handle them firmly and take steps to minimize any kicking and thrashing.

As with most prey species, a rabbit's eyes are spaced far apart on its head, giving it a wide field of vision but leading to a blind spot under its nose. The rabbit has a very dexterous, very sensitive split lip, and it depends on that and its whiskers to find its food.

The ears are perhaps the rabbits most distinctive feature, and because they are so highly vascularized, critical in heat regulation. Rabbits do not sweat, and must depend on passive thermoregulation to cool themselves. Rabbits pant to exchange body heat with the atmosphere, and blood vessels in their ears expand, so their ears function much like radiators to exchange body heat with cooler air. This is also why good ventilation in summer is critical for rabbit health, but more on that in Chapter 6.

Rabbit fertility (how easily they become pregnant) and fecundity (how easily they carry the pregnancy and deliver, and how often) are legendary. Consider that it is possible for a doe to give birth to a litter of 6–12 kits between six and eight times a year. As you might imagine, larger litters will take a toll on a doe's resources, and the larger the litter, the less frequently she will be able to produce them.

Rabbits have two separate uterine horns, each with its own cervix. It's possible for a doe to have kits in one horn and not the other, but much more commonly they will have kits in each horn, and one horn will give birth and then the other.

And one other thing that makes rabbits distinctive in the mammalian kingdom is the behavior of *cophrophagy* (the practice of consuming one's own feces). It sounds disgusting to humans, but as we will see in Chapter 8, it makes perfect biological sense to the rabbit.[4]

So there's a quick overview of some of the things that set rabbits apart (and probably more than you wanted to know about taxonomy) from other livestock and can make them a challenge to manage. But with good husbandry, and a little awareness of those factors making domestic rabbits unique, you will be well on your way to running a successful rabbitry.

CHAPTER 4

Rare Breeds and Conservation Breeding

Rare Hare's Story

Our rabbit journey started innocently enough. After news of the umpteenth food recall in a relatively short period of time in 2002, Eric and I decided we needed to take some direct control over the sources of our food. After rehoming a pair of New Zealand rabbits, Eric stated that he had raised rabbits before. We could probably handle raising a few rabbits…just for ourselves, you know?

I had never eaten rabbit, but I'd never run across much in the way of food I wasn't willing to try, so we soon acquired some odds and ends of what was available locally, including a couple of Satins.

But at the same time, we were becoming much more active and interested in the work of The Livestock Conservancy (TLC—formerly the American Livestock Breeds Conservancy),[1] and we started looking into breeds of rabbit that were becoming rare.

In 2003, the American Rabbit Breeders Association (ARBA)[2] national convention and show came to Wichita. Armed with enough information to be dangerous, I went to the show, interested in seeing some of the breeds I'd been reading about. (If you've never been to ARBA nationals, it's impressive to see thousands of rabbits and guinea pigs in one place.)

But, in all the rows and rows of coops, I couldn't easily find the rabbits I was seeking. Or really, hardly at all. Finally, I stubbornly located five Americans over in a corner almost under a stairwell of the

FIGURE 4.1. The Wichita 2003 ARBA Show: 20,000+ rabbits and cavies (guinea pigs) on exhibit in one room. Out of the frame is the corner where the American Blues and Whites were stashed.

exhibition hall. A couple of whites, and three blues, if I remember correctly. There were also one or two Hotots, and if I saw American Chins or Silver Foxes I don't remember them; I had specifically gone to find the Americans. They were gorgeous, but seeing so few in all the hundreds and hundreds of representatives of other breeds was pretty sad. The vast majority of rabbits on exhibit were small pet types. Which are fine, but not what we were looking for.

So, to make a long story short, not being able to find what I want makes me more determined to find it, so a quest was on. Email was still relatively new then, and Facebook wasn't even an idea, so searching involved actually calling people and doing a little old-school detective work.

Finally locating breeders in Indiana, we had to wait a while for available breeding stock, but eventually a road trip happened, and we came home with a car full of rabbits of various breeds.

The rest, as they say, was history.

The original animals were beautiful stock, but show rabbits aren't selected for the same traits as production rabbits, and production was of primary importance to Eric. Everything on the farm needs to pull its weight, by providing either food or income.

As Eric began breeding and selecting, one of the initial challenges we faced was what to do with the surplus and the culls. One home freezer can only hold so much.

So the meat production side of Rare Hare began as a result of our conservation effort, rather than the other way around. It has expanded to a level we would originally never have conceived, but the heart and soul of the farm has always been rooted in rare breed conservation, meaning breeding animals close to the breed standard and the production levels of their heyday in the early 20th century.

The Livestock Conservancy added rabbits to their mission in 2005. Rabbits have always been a little different for TLC to categorize on their priority list.[3] Other species (such as cattle, sheep, and horses) are categorized by annual registration numbers. Registration for these species involves submitting an application, and the animal is registered based on its pedigree.

Registration for rabbits is a bit different. Rather than submitting an application, to be registered with ARBA a rabbit has to be inspected by a trained inspector, pedigree has to be verified and when all is in order, a tattoo placed in the right ear of the rabbit. (This is why individual identification tattoos are always placed in the left ear.) Registration of rabbits is often done at rabbit shows. Since there are obviously rabbits out there that never get registered, numbers are calculated and supported by data supplied by breed clubs.

Conservation Breeding

So what exactly is conservation breeding? For us, it meant adhering to the criteria and production traits that made rabbit one of the most widely available and popular animals back in the early 1920s, when people depended on them for food or their livelihood.

If you are interested in breeding rabbits, you can see why it is important to set your own goals and criteria for selection. It's fine to have a goal change and evolve over time. Initially your goal might just be to have animals that can reproduce regularly, and then once that's achieved, you can move into selecting for other traits such as color, meat quality, or fur.

Each mating is important. And especially if you are doing conservation work, detailed records of those matings are important. Without good records, it simply isn't possible to track inbreeding, or evaluate which animals bring the best qualities and pass them on to their offspring. And, when availability of new bloodlines is limited, a judicious use of several breeding strategies is necessary.

FIGURE 4.2. A uniform group of Silver Fox rabbits; while their accidental uniformity of pose is amusing, this is also a very consistent group of does which are potential breeding stock.

Breeding Strategies

Inbreeding

The old adage is "It's linebreeding if it works, inbreeding if it doesn't." This is both highly accurate, and an oversimplification.

Both inbreeding and linebreeding involve mating animals that are more closely related than the average population. Inbreeding has been given a bad rap, and rightly so in many cases. But without inbreeding, it would be difficult to concentrate positive traits with any sort of consistency.

Inbreeding is defined as the selection of animals for mating that are more closely related than the average of the population. There are shades of severity in inbreeding. A brother–sister mating between two animals whose parents do not share any common ancestors might be more successful than a brother–sister mating between more closely related individuals. In other words, the more common ancestors found in a pedigree, the more the variability of the genetic material is reduced.

Signs of Inbreeding Depression

- small litter size
- failure to thrive
- behavior issues
- regression in positive qualities like size or uniformity
- appearance of undesirable traits

Inbreeding concentrates genes, both good and bad. It varies from breed to breed and individual to individual how much can be tolerated before problems arise.

Linebreeding

Linebreeding involves mating animals that share some of the same genetic material, but not to the extent of full sibs or parent–offspring. Eligible mates include grandparents bred back to an individual, half-sibs mated together, and other instances where common ancestors are a generation or two back in the pedigree.

Very few breeds, if any, were created without some form of line-breeding. Even Landrace breeds, by virtue of the selection pressures the environment places on them, will concentrate the genes of a few hardy animals that thrive and gain reproductive advantage over their peers.

Pedigrees and good reproductive records are critically important to breeders. Knowing a goal at the outset will also help to make sure you are aware when things start to slip, and you can at least know when a correction is needed.

There are no hard-and-fast rules governing what concentration of a particular individual is too much, or not enough. Expert breeders who are not afraid to cull heavily can make good use of both inbreeding and linebreeding in their selection process. The key is to know what your goals are, know what you are hoping to accomplish, and know when too much is too much.

Outcrossing

Outcrossing, conversely, is the mating of animals that are less closely related than the average. This can be accomplished by combining lines from different rabbitries, or different and isolated bloodlines within the same rabbitry. Most rabbitries, just by virtue of isolation, will practice some form of linebreeding, selecting for traits they find desirable, and concentrating those whenever possible.

Each time new stock is added, or a new bloodline is introduced, all the genes have to sort themselves out again. Depending on what criteria the other breeder has selected, you may experience a dip in quality or quantity of some of the traits you intend to select. Until you have a few litters and a couple of generations to evaluate, you will not know exactly how all genetic material will sort itself out.

Outcrossing can improve some positive traits, but a herd of rabbits that has been selected for qualities such as litter size, rate of gain, and meat quality will likely experience some resorting of genetic material when its breeder outcrosses the stock.

Outcrossing is necessary at times to prevent inbreeding depression, or the concentration of undesirable traits. But just be aware when doing so that you may plateau on your own selection for a while.

When selecting initial stock, novice breeders are better off to stick with animals from one herd, and purchase breeding stock from that one herd that has as much genetic variability from within that group as possible.

Crossbreeding

Crossbreeding is breeding individuals from two separate and distinct breeds together. Because of the differences in the genetic makeup between breeds, crossbred individuals often (but not always) possess the best traits of each breed. This effect is also often referred to as *hybrid vigor*, although this is a bit of a misnomer, as a true hybrid would involve crossing two distinct species. (Think of the mule, a cross between a donkey and a horse.) The initial cross between breeds might be highly uniform in appearance and traits, but breeding these crossbred individuals together might result in a completely mixed bag of traits, as the genes sort themselves out.

Crossbreeding is best used for breeding animals that will directly go to harvest, and not be retained in the herd. If you'd love to experiment with breeding and see what happens, crossbreeding can be a lot of fun, but if your goal is consistency you might find yourself frustrated by random traits popping up here and there.

None of the breeding strategies described in this chapter are in and of themselves wholly bad or wholly good. Each has a place in breeding, and their use depends on both the makeup of the original stock and their use by the individual breeder. The important things are to have your own goal and to collect as much information as possible to inform your decisions. With rabbits, it's possible to see results in a shorter period of time than with cattle, sheep, and goats, but it will still take time and good records to see and track improvement.

Variety, Line/Family, and Type

Among the individuals of a breed, there are often variations in surface characteristics such as color. This phenomenon is called *variety*. Variety is why there are both blue American rabbits, and white American rabbits, as well as the other array of color varieties seen in many of the breeds more common on the show table. The general genetic makeup of the rabbit lends itself to a whole bunch of color combinations and patterns.

A *line*, or a *family* is a further subgroup of a variety. Animals in a line or family are often descended from one sire, one bloodline, or from a specific breeder.

A *type* often represents a subset of physical characteristics, such as meat, or fur. (Think of how dairy cattle differ from beef cattle in appearance; that is a clear difference in type.) A meat rabbit will have immediately visible characteristics when compared with a show type, or a wool type.

Type distinction is more common with other species of livestock that have characteristics of dairy or meat or fur, but when a person speaks of a meat type rabbit, its physical characteristics should be immediately apparent.

Starting with a Trio or Two Pairs?

When selecting animals, most people begin with a trio (one buck and two does). This is the most common purchase; however, within one generation, you are at a genetic dead end.

Sire Influence
One buck can service at least ten does. We keep more than that on hand, as we maintain several individual lines within the different breeds. You will need more males than average, especially if you are doing conservation breeding and maintaining several different bloodlines.

Unless you live close to the breeder and have the option of returning for new stock, it is a better strategy to purchase two pairs of breeding stock, the two bucks being as unrelated as possible within the confines of the breeder's particular herd. Two bucks can be used alternately on each other's daughters before much inbreeding is noted. At that time, the novice will have a lot more experience in selecting and evaluating stock, and an outcross will not throw them off too much.

Having two bucks also gives you a backup. Sometimes rabbits just die, for no apparent reason and with little to no warning. If you only have one buck, and haven't got your does bred yet, if he dies you have a problem. Having a second buck can at least keep you in the breeding game.

Heritability of Production Traits

Production traits can loosely be defined as the qualities that an animal possesses that allow it to raise offspring easily and produce more young in a cost-effective manner.

Heritability is the ability, or the ease, with which certain traits are passed on to the next generation. Traits that are highly heritable are passed on readily, and their influence can be seen in a generation or two. Traits that are lowly heritable take much longer to assert their influence. It's not that selection is impossible, but just that certain traits will be much slower to respond to selection.

As in most species of livestock, traits such as reproductive ability, litter size, and disease resistance are lowly heritable. In other words, it is not easy to increase the average litter size in your rabbitry by only keeping individuals born to does with large litters. Nor does it mean that the female offspring of a doe that only had two or three kits should automatically be culled. Those offspring may not have litters of 12 or more, but they might slowly begin to improve the litter size of that line.

In our experience, a better selection strategy is to have a "floor" for the entire doe herd. For example, if you determine that it takes at least six kits weaned per litter for the rabbitry to be profitable, then no doe with less than a six-weaned-per-litter average will be retained. This allows each doe selected to show what she's capable of producing, and over time, the average number weaned for the entire herd will rise. It should also be noted that good management will often have more impact on reproductive qualities than selecting individuals for lowly heritable reproductive traits.

Other qualities such as growth and feed efficiency are considered moderately heritable. These traits will respond better to selection than lowly heritable traits.

It's going to be difficult for the average (and by average I mean one without a huge pool of stock to choose from) rabbit breeder to actively select for traits such as feed efficiency without a lot of data tracking, but what works for us is to select the largest rabbits out of a litter, and as long as they meet the rest of the selection criteria, put those into the breeding barn. Over time, the difference between the bottom and the top end of the litter will shrink. Of course, management will have a huge impact on whether or not the rabbits are able to express their full genetic potential. If a pen is crowded and there is a lot of competition at the feeder, it may be best to split them up and see if any outstanding individuals emerge with less competition. On the other hand, if in that crowded pen there is one that completely leaves their pen mates behind, that rabbit might be a shoo-in for a role in the breeding barn.

Carcass traits, however are much more heritable. These traits are easily influenced by selecting an individual with superior carcass qualities and using those offspring for breeding.

A good, meaty rabbit that has lots of substance and flesh will pass that on pretty readily to their offspring.[4]

A word to the wise though: Don't become focused on one overall trait to the exclusion of all the rest. Focus on overall balance in the animals you select, rather than searching for that one outstanding individual. You will be able to make many more consistent strides in your genetic selection by focusing on good animals with a mix of good traits rather than excluding animals that aren't "perfect." And perfect is in quotation marks because there is no such thing as a perfect animal of any species.

When selecting rabbits within your own herd, be objective. No litter is ever made up entirely of animals that are good enough to be breeding stock. After a few generations of learning and selecting, the differences between the top and bottom animals will shrink, but it will almost always be there, just more subtle. It might be a half pound difference in weight at a certain age, or a degree of shade in color, but it will be there.

Another strategy is to base selection on an individual animal's net merit. A doe with a decent litter size and decent rate of gain might be a better overall pick than one doe with poor litter size and good rate of gain, or a doe with good litter size and poor rate of gain.

Don't be dazzled by the animal that is the Supreme Champion in the show ring, either. Often the second or third place animals are very close to that top rabbit in quality, and might be available at a lesser cost.

That champion may not necessarily pass on those champion qualities to his offspring. The animal that came in second place is still overall a pretty good animal, and may pass on a more consistent genetic package.

Some Sage Advice

In poultry breeding, there is an old adage known colloquially as *The Law of Ten*.

This adage basically says that in order to find 1 good individual, 10 must be produced; a great individual requires 100, and the exceptional individual, 1,000.

This isn't a rule in the strict sense of the word, but it does highlight the work and dedication it takes to improve a breed of livestock, any breed. Strict culling and adherence to breed standards must be kept. Anecdotally, we've found that about 10 percent of the animals we produce will be good enough to be considered as breeding stock. This wasn't set forth as a goal, but in looking over years' worth of data, it proved true that only 1 in 10 had what it took.

Another old adage is "whatever you tolerate in your barn, that is what you will have." In other words, if you do not select against poor-quality animals, it will be difficult to improve your herd.

In the early stages of learning to select, it can be difficult to see much difference between your animals. But as your eye for selection develops more discernment, the subtle differences will become more readily apparent.

Selection criteria can be a double-edged sword in rare livestock breeding.

On one hand, people sometimes get the idea that because it is rare, every animal must be considered as breeding stock. And while there are some cases where this is necessary (in the case of breeds which have become so low in numbers that their survival is threatened), for the most part it is beneficial to the population to make good selection decisions. The cases of breeds which are so rare or endangered that every animal must be considered as a breeder have thankfully become more rare themselves.

And always, mathematically, you will have more surplus males.

It often takes large numbers of offspring before you can make any sort of a significant change to the genetic makeup of your herd. This can be accomplished over time, by a few litters a year over the course of many years, or by a large number of litters over a shorter time. Either strategy involves attention to detail, and strong culling. Be prepared to cull frequently, and cull hard, even with rare breeds. Evaluate males critically, and retain the best of the best to improve your herd.

CHAPTER 5

Selecting and Handling Rabbits

Choosing the Rabbits

If possible, purchase two unrelated bucks, and two or four as-unrelated-as-possible does. Your bucks should be as diversely related as possible. This will allow you some time before your herd becomes so interbred that the rabbits suffer inbreeding depression.

It's not necessary to get rabbits from multiple breeders, at least not in the beginning. A single breeder with several years of experience will have developed a group of rabbits that work and are productive for them, and will have done a lot of selecting before they ever get to the point of offering them for sale.

It needn't be someone with decades of experience (although that's wonderful), but try to find someone who has worked with the breed for at least a couple of years. That will have given them enough time to be familiar with the traits and quirks of the breed, and they will have had a few litters and a few breeding seasons under their belt. Not to say someone new to the breed will not have good stock, but experience takes time.

If you find a breeder who will let you pick out your rabbits, make sure you are familiar with the breed standard and know what to look for. More information is found in the "Qualities for Selection" section later in this chapter.

Request Pedigrees

Most breeders keep pedigrees on their stock; only occasionally will you find one that does not. Without a pedigree, or information on their breeding history, it will be up to you to find out how related your new rabbits are and figure out how to proceed with your own breeding program.

FIGURE 5.1. These jugs of water and bags of food are ready to accompany two trios of American Chinchilla to their new home.

Bringing Your Rabbits Home

Whenever possible, visit the breeder to pick up your rabbits yourself. This gives you an opportunity to meet the breeder, and if possible, tour their rabbitry and pick up any tidbits of information that they can offer. Many more experienced breeders are a wealth of information not only on their particular breed, but also on details about good rabbit husbandry. Most breeders will be happy to show you around, but some will not. This doesn't mean that they are hiding something; it may be that they have litters due and prefer to keep activity to a minimum, or that they don't want to risk possibly bringing in outside bacteria.

Consider also that you will be taking your rabbits into an entirely new environment. Anything you can do to make the transition smooth will help them settle in and be as stress-free as possible, so note details about their home rabbitry: how the feeders are set up and how the watering system functions, how the ventilation works as well as what time of day the chores are done.

When you arrive home with your new rabbits, plan on quarantining them from any rabbits you already have for at least two weeks. This will allow you to observe them for any health issues, and give them an opportunity to adjust to your schedule.

Ask the breeder you got your rabbits from to give you a few days' worth of feed and water to take home with you. This is not a common practice, so don't be surprised if you get a raised eyebrow when you ask. But because it's relatively easy for a rabbit's digestive system to be disrupted, it pays to make the effort.

Switching them over to your feed slowly will help minimize stress during the transition. It is not unusual for rabbits to go off feed initially when adjusting to a new environment; having their familiar feed will help ease the stress.

Water is often overlooked in this process, but water smell and taste can vary widely from location to location, and while an animal can survive for several days without eating, dehydration occurs quickly and the effects can be devastating. A two-liter pop bottle full of water from the breeder can save a lot of heartache.

> When we send out breeding stock, we always send a baggie with at least one weeks' worth of feed, and a gallon (4 L) jug of water to ease the transition. For the first three days, we ask people to give only our feed and water, and then after the third day, begin mixing our feed and what they will be feeding at a half-and-half ratio until that runs out, and then switch solely to what they will be feeding from that point on. The same with water.

Handling Rabbits

Rabbits are both very hardy and very delicate. With their light skeletal system, they can suffer fractures easily if mishandled, but I've also seen small kits take an accidental tumble out of a pen door and suffer nothing worse than a little bit of "whoa, what a ride!"

Large/Adult Rabbits

The powerful back legs of adult rabbits can generate enough force, if applied at a wrong angle, to break their spines, usually right above the pelvis. This, to put it bluntly, sucks. A rabbit that you have invested enough in to get to the adult stage is a major loss. Rabbits rarely recover from such an injury, and the most humane thing to do is to euthanize them and put them in the freezer right away. On the rarest of occasions, a forceful kick can just cause a bruise and not a complete severing of the spinal cord. You can give the rabbit some time to see if this is the case. But be aware that a rabbit with a broken back cannot engage in normal eating behaviors, including cophrophagy, and it will be just a matter of time before it becomes unhealthy. Not being able to clean itself will also lead to *flystrike*, which is a problem you don't want to put your rabbits through.[1]

Handle rabbits carefully, but not timidly. Because rabbits are prey animals, they can easily pick up on your hesitancy, which will cause them to become concerned and potentially become agitated. Also, if they don't feel securely and firmly held, they will try to escape. And, some rabbits just don't like to be held, and would rather scratch you than go along for the ride.

Rabbits, like domestic cats, have a very loose scruff around their neck. When picking up a rabbit, get a good handful of that scruff, and lift them firmly, but gently. Some rabbits will kick or try to wriggle around and kick aggressively; some will quietly just hang there.

If you have the quiet kind of rabbit, you can carry them for short distances by that scruff. However, this takes a pretty good measure of hand strength. If you aren't sure you can get from point A to point B without losing your grip, don't chance it.

If your rabbit is not interested in complying quietly, as you pick it up, tuck it under your arm, and carry it in the same way a good wide receiver should carry the football. The hindquarters can be cupped in your hand, the rabbit cradled against your body, and its head tucked under your elbow.

It is still possible for a rabbit to scratch while it is tucked against you, although keeping them close to your body reduces the potential. It's always a good idea to wear a heavy shirt or denim jacket when handling your rabbits. Bare skin and rabbit claws are not a winning combination.

Rabbits that are handled well will most times get used to it and do not struggle too much at being moved from place to place.

If you have a rabbit that, upon opening the pen, proceeds to run around the pen stamping and protesting, it is going to be

FIGURE 5.2. Eric has this American Chinchilla tucked securely, and the rabbit is relaxed and not struggling.

difficult to try to grab the animal without possibly injuring them by grabbing a leg, or something else incorrectly. Slide your hand and arm into the pen, and position it so that when the rabbit makes a lap around the pen, you can close your arm around it, and pin it against the pen front. Block the pen door with your body. You should then have the rabbit's head in the crook of your elbow, and you can reach in with the other hand and grab the rabbit by the scruff of the neck. Be ready as you pick it up to close it against your body as soon as the rabbit is clear of the pen door.

FIGURE 5.3. Eric has captured this rabbit against the side of the pen and can reach in and grab its scruff without difficulty now.

Occasionally you will have a rabbit that is downright nasty about being handled, and will attempt (or succeed!) at biting you when you put your hand in to pick it up. This nastiness is different from a doe that might be protective of her litter, or pregnant and cranky. Rabbits that have a nasty disposition should be on the short list for freezer camp. A rabbit that is actively trying to bite you can be distracted by one hand put in the cage to focus on, while at the same time you grab the scruff with the other. This takes good timing, a lack of hesitancy, and a glove on the hand you use to distract the rabbit. When rabbits bite, they can clamp down and not let go. If they have your skin in their teeth, resist the urge to pull away if possible, since this can cause more damage. Call for backup. I had a particularly nasty rabbit get ahold of my upper arm once, and I had to wait for Eric to come help me disengage his teeth from my flesh. He went to freezer camp (the rabbit, not Eric), and I still have a scar. But not nearly what I would have if I had had to pull him off me. At the very least, that rabbit was tasty to eat.

FIGURE 5.4. Carrying a rabbit for any length of time this way does take a bit of grip strength, which Eric definitely has. For me, this grip works best for moving rabbits from pen to pen or shorter distances.

Small Rabbits/Kits

Smaller rabbits can be held by the scruff, like the adults, or by the hips. The hip carry is often much easier for the handler and less stressful for the rabbit. Youngsters don't always have as much loose skin to get ahold of, and they can wriggle an incredible amount. The hip carry won't hurt them a bit, and they tend not to struggle when carried this way most of the time. Another advantage is that you will have control over those hind legs.

Sexing Kits

Sexing kits is a basic skill that can be a little challenging at first, but it will save you a lot of time and effort in the the long run. Dividing your litters by gender will give you a lot better chance to

- ▸ evaluate them as potential breeding stock
- ▸ let them get to a good weight to process without running the risk of fighting
- ▸ rest assured that young does don't get bred by siblings

We explain other reasons to split up litters in Chapter 10.

It's hard to see kits' genitalia when they are small, but it's much, much easier to handle them at that point, so practice trying to sex them from about four weeks old. Recheck at two-week intervals until you are confident you have made the right determination.

Grasping each kit by the scruff, turn it into your palm to support and control its wriggly little movements. Using the thumb and forefinger of the other hand, press between its legs until you see the genitals.

FIGURE 5.5. (*left*) Buck kits have an opening which looks like a circle. This is the penile sheath. In older bucks the testicles can be quite easily seen, and often when sexing an older buck, the penis will protrude.

FIGURE 5.6. (*right*) Doe kits have more oblong or oval openings. This is the vulva.

A male will have an opening that looks like a circle, a female's will be more oval. The younger the kits are, the more similar they appear, but with practice you will be able to distinguish shape quickly and easily.

Qualities for Selection

When dealing with rare breeds, it can be tempting to think of every animal in a new litter of rabbits as potential breeding stock. However, especially in a rare breed, rigorous selection standards need to be applied to any potential breeding animal. In all but the most critical of cases, rare breeds are not served by keeping mediocre genetics in favor of increasing numbers.

Each breed has a standard which can be found in the American Rabbit Breeders Association (ARBA) Standard of Perfection (SOP).[2] This publication details the requirements for each breed including Junior and Senior weights, body type, fur quality, color varieties, and faults or disqualifications. Even if you are raising rabbits solely for your own consumption and never plan to show or sell breeding stock, adhering to the breed standard in selecting replacement stock is vital; otherwise a commercial mixed-breed rabbit would meet your needs just as well.

The SOP includes how much importance is given to each quality of the rabbit; for example the Silver Fox and the American Chinchilla were both developed as dual-purpose breeds for both meat and fur, and the quality and coloration of the fur are half of what these rabbits are judged on. The SOP is published every five years and is a reasonably priced investment for a breeding program.

That being said, there are certain general qualities that a breeding stock rabbit should possess.

FIGURE 5.7. This rabbit's teeth were allowed to get overly long for the photograph. Afterwards, his teeth were trimmed with a set of piglet tooth nippers, and he went to the fryer barn.

Head

Examine the mouth and teeth. The rabbit should have top incisors that slightly overlap the bottom ones. This allows for the rabbit's teeth, which grow continuously throughout its life, to wear evenly. *Malocclusion* (misalignment between upper and lower teeth) is generally considered to be heritable, and most breeders cull rabbits with this condition. However, malocclusion can also be the result of injury from biting and pulling on cage wire or feeders. Before discarding a promising line of rabbits, or one that considerable investment has gone into acquiring, you can test whether or not any malocclusion present was inherited by inbreeding (for example, breeding father to daughter) to discover if the trait is hereditary. If the results are unfavorable, then be prepared to cull. Maloccluded rabbits will eventually have problems eating, and their physical condition will deteriorate.

The rabbit's eyes should be clear and bright, free from discharge or cloudiness. The nose and mouth should also be clean and clear. White nasal discharge is a symptom of *snuffles*, a highly contagious respiratory infection.

FIGURE 5.8. This is a good American White doe; she met all the selection criteria and went to the breeding stock barn.

Lopped or droopy ears disqualify any breed that does not specifically call for lop ears. Overtly short ears are also a fault in most breeds, and with good reason, as the ear serves as a radiator and allows the rabbit to cool itself in warm temperatures.

Body

When evaluating the rabbit, allow it to pose naturally on the table; don't try to scrunch it up. The body in general should be symmetrical, with good wide shoulders that have adequate space between the shoulder blades. A good rule of thumb is that you should be able to slip at least one finger, minimum, between the shoulder blades of the rabbit. The body should taper slightly from the hindquarters to the shoulders, but the shoulders should not be pinched in. The entire rabbit should have a firm, meaty feel—solid throughout.

Legs and Feet

The feet should be strong and well centered under the body. Look on the soles of the hind feet for any sign of sore hocks. *Sore hocks* can be caused by environmental factors, i.e., rough wire flooring, but often it is a structural fault in which the rabbit's feet do not carry its body squarely.

Weight

Each breed has an ideal weight for bucks and does as well as a range that is acceptable. Barring environmental issues, rabbits that do not meet at least the minimum weight requirement should not be kept. Allowances may also need to be met for does that have just weaned a litter, but in general a good doe will not lose much condition during kindling and lactation. One of the soundest investments a new breeder can make is to purchase a good scale.

CHAPTER 6

Housing and Environment

The Barn

If you are just starting out, it is perfectly acceptable to make use of existing structures for your initial rabbit building. After a little time and experience, you will know if rabbits are something you wish to continue in, and at that point you can decide to create and build your dream barn. If rabbits turn out not to be the species for you, you are only out a little cash and your time.

With that said, there are certain criteria for any rabbit facility, and if those are not met, your efforts will likely result in frustration, unnecessary expense, and a poor quality of life for your rabbits.

Environment

The rabbits' environment is probably one of the most overlooked—but critical—aspects of running a successful rabbitry.

Environment is more than just the temperature or the weather, although those are important. Environment refers to the overall quality of care and attention to the needs and natural behaviors of the rabbit.

A rabbit that is continually under stress from its environment will be, for example, less likely to conceive or have a lowered resistance to disease.

Take some time before starting out to evaluate your situation as it is. Take into consideration the climate where you live: Is it normally hot? Cold? Humid? These factors will have a huge impact on your success or your challenges.

Rabbits tolerate cold weather much better than hot for extended periods of time. And they will always struggle in a highly humid environment, no matter the temperature. In winter, many buildings are set up more for human comfort than for the rabbit's comfort. They lack adequate air movement to keep humidity, and therefore ammonia, at a minimum.

Keep in mind that a long, narrow building will be easier to ventilate than a more square one, and that ventilation is critical in the summer for cooling, and in the winter for air quality.

Noise

When selecting the area for your rabbit housing, take some time to look around and note anything that might be a potential source of stress for rabbits. Dogs barking, children running around shrieking, loud machinery, and other such noises can create an overly stressful environment for rabbits, and can short-circuit all other attempts to be successful.

Rabbits have been hardwired over millennium to panic first; they really never get around to asking questions later. A constant barrage of stimuli can convince your doe that having kits and raising them isn't even close to being in her best interests. It can cause her to stress so much that she fails to conceive, or even if she does, she may be so stressed that she cannot take proper care of her litter.

Keep pets such as dogs and cats away from your rabbits. Even the most well-behaved and well-trained dog can cause a rabbit undue stress even if the rabbit just sees or smells it. By all means, involve your children in your rabbit venture, but make sure they understand why they shouldn't make loud noises or sudden movements, and that they respect these rules.

If you have your rabbits set up near where noise is unavoidable, put an inexpensive radio in the barn. A constant low level of noise can mask a lot of sudden noises, and provide auditory consistency that rabbits will find comforting.

Sanitation

You wouldn't want to live in a room with feces packed in the corners. Neither do your rabbits. Set things up so your cages can be cleaned easily, and keep manure from building up in corners or on resting boards. Manure can attract flies, which can cause *flystrike* (flies lay eggs in the rabbits' fur or any open wounds, which leads to a maggot infestation, which can be extremely hard to treat and is often fatal for the rabbits).

Take the time to disinfect your pens between litters. This can get rid of bacteria and other bad stuff; it's well worth the trouble it takes. A spray bottle with bleach water can go a long way to keeping things clean. If you have many rabbit pens, a pump sprayer is a good investment,

---

About Fly Control

It's impossible to completely eliminate flies from the outdoors. But a couple of steps taken at the right time can help keep their annoying and disease-transmitting presence to a minimum.

The fly traps that you can purchase at the farm supply store work best if you can get them out early in the spring, just as things start to warm up good, and before you really start to see flies. This enables the trap to catch and eliminate the breeders, and can exponentially reduce the number of flies produced for the year. Put the traps away from your barn in a shaded area, out of the wind. Flies don't like sunny, breezy areas and a good shaded location will help draw them to the trap.

Fly tape and hanging fly strips also work well, but you must take care to change them out once they become full of flies. Tape and strips are also notorious for catching fur. And again, the earlier they get put out the better.

Flies also love a good water source. They will gravitate to open water crocks, or open jars, and any wet places on the floor. And they love to lay eggs in a damp environment.

FIGURE 6.1. Because of the sheer number of pens we disinfect on a regular basis, we invested in the larger pump sprayer, and a handheld torch with the igniter on a hose, rather than one piece. This saves wear and tear on the wrist when using, and is much less awkward to handle on a regular basis.

FIGURE 6.2. Sunlight is one of the most powerful disinfectants available.

otherwise a trigger spray bottle will do the trick. An outdoor grill cleaning brush works wonders for removing feces and other solid material from the pen. Another easy way to disinfect an empty pen is to use a small propane torch, and run the flame over the wire. This removes hair and cobwebs, also. Of course, be careful not to hold the flame in one place too long, use a quick back and forth movement. This technique even allows you to run the torch over the plastic automatic water lines.

And always keep manure cleaned away from under cages and your rabbits.

Disinfection

At some point it will be beneficial to disinfect your rabbitry and equipment.

One of the cheapest and best natural disinfectants is sunlight. UV light will kill most pathogens. It can be a spring and summer project to bring all the equipment that can easily be brought outside (such as nest boxes not in use, feeders, drinkers, tools) and lay them out in direct sunlight for an afternoon.

For the pieces such as the pens themselves, spray a mild solution of bleach water on the cage surfaces. Chlorine bleach is relatively inexpensive and readily available.

For heavy-duty cleaning, there are other commercially available disinfectants, but be sure to rinse adequately after using them, as harsh residues can be irritating to the rabbits' feet, and fumes can irritate their lungs.

Housing and Environment   45

Environment and the Coat of Death

I have one coat that I no longer wear in the barn. It's the Coat of Death, a black, heavy nylon windbreaker, and on the couple of occasions I've forgotten and worn it into the barn, the rabbits have reminded me they don't appreciate it one bit.

Our rabbits are used to people moving around; Eric moves around in the barn many times a day, in and out, over and over, and the rabbits accept him without qualm. I am not in and out nearly as much as Eric is, but most of the time it is fine…except when I'm wearing the Coat of Death. There is something about this black nylon coat (probably combined with me moving too fast and not paying attention to what I'm doing) that invariably spooks the rabbits. Whether it's the sound it makes, whether the black color reminds them on an instinctual level of the shadow of a giant bird of prey, or all of that combined with me moving much differently than Eric does, causes panic in otherwise calm but alert rabbits. It's a quick way to spook a rabbit so much that they can freak out and badly injure themselves.

I tell you that story not to keep all black windbreakers out of rabbit barns, but rather to illustrate that what might be a tiny insignificant thing to you could register on an entirely different level for the rabbits. Take a moment to evaluate the place your rabbits will live from the rabbits' perspective, not yours.

Dish soap also can be used, but it has to sit on any soiled area for a while to have much effect. The advantage is that it is cheap and readily available as well.

All disinfectants will perform better if there is no fecal material or dirt on the surface of the thing being disinfected. Take a few moments to rinse with plain water, scrape to remove poop, or do whatever is necessary to enable a disinfectant to do a good job. Cleaning in advance ensures that your disinfecting efforts won't be wasted.

Biosecurity

Biosecurity doesn't necessarily mean wearing a hazmat suit or spraying yourself with disinfectant every time you handle a rabbit.

It means being careful with bringing new stock into your rabbitry, limiting the number of visitors from other farms, and keeping pests, rodents, and predators at bay.

It also means making sure pens—and the barns themselves—aren't overcrowded. And it also refers to daily observation of herd health and cleanliness.

Waste Management

Anything that eats, poops. It's a basic fact of biology.

A 10- to 12-pound (4.5–5.4 kg) doe and the 28 offspring she can produce in one year will yield about 6 cubic feet (0.17 m³) of manure. A single herd buck or doe without a litter can produce 3 cubic feet (0.08 m³). And that's without any bedding or nest box material mixed in.

No matter what sort of rabbit facility you design, you are going to have to remove waste. Manure that builds up in the barn will become a haven for flies and other insects, which can carry disease and are just plain nasty. Ammonia evaporating from urine can build up and negatively impact air quality. To keep your waste pile from becoming a nuisance, you will want to place it a distance from your rabbit barns.

Consider carefully what your plan will be for removing waste. If you are planning to renovate an existing structure for rabbits, consider how easy it will be to get the manure out of the barn. It makes cleaning day longer

FIGURE 6.3. This handy cart allows dumping of waste directly into the manure pile. I don't even want to guess how many tons of rabbit poop this thing has hauled over the years.

if you have to walk a shovel from one end of the barn to the door. Will the door open wide enough to allow a wheelbarrow or small dump cart inside? Is there a door that can be opened wide enough to pull a tractor with a loader bucket up to it? Or pull up a short distance away so that you can dump the wheelbarrow into the bucket?

And more importantly, who is going to be the manure jockey?

There's no doubt that waste management is one of the least pleasant parts of raising rabbits, but it's one that cannot be skimped on, for the sake of the rabbits' health.

Set up a waste management system that works for you. You can hang

FIGURE 6.4. We affectionately named our manure pile Mount Shista. Eric turns it regularly with the tractor. A local greenhouse buys a couple of truckloads of it a year, liking the fact this manure is free of weed or straw seeds.

tarps under cages and funnel the waste into a bucket to be dumped later, create an automatic scraping system and so on. But at some point or another you are going to have to handle that waste. Whether you scoop it up and put it in a wheelbarrow or have it flow into buckets, make it easy to handle.

Barn lime can also be a useful product. Get in the habit of sprinkling lime on the inevitable wet places under the cages where urine drains. This will help break down the ammonia, dry out the wet places, and kill off any fly larvae.

There are as many ways to handle the waste as there are clever people to design them. Whatever you choose to do, keep a few things in mind:

1. Manure draws flies. Keep that pile away from your rabbit barn.
2. Whatever substrate you use, make sure it is porous to allow urine to percolate through. Cement may be easier to clean, but it will

become wet and laden with ammonia, increasing the humidity in the barn and making the air quality very poor.

3. Rabbit manure is fantastic for gardens; if you haven't taken up gardening yet, this is the perfect opportunity.

Rabbit Manure for Gardens

Fortunately, rabbit manure is pure gold if you garden. Rabbit manure is one of the few types of manure that can be applied directly onto plants and soil without going through the compost pile. It is high in nitrogen, phosphorous, and potassium but low in acid so it does not burn plants. I've even put it directly on some very finicky houseplants and was rewarded with something just one step short of *Little Shop of Horrors* quality. (Now if I could just train that plant to go after mice....)

Rabbits fed a pelleted diet will produce manure that is relatively free of noxious weed seeds. Rabbit manure is safe enough to be used at the rate of 2 pounds (1 kg) per square foot in the garden, although several applications are better than one or two large ones.

If you have rabbits and don't garden, perhaps you should consider it. You are going to have manure to dispose of, so you might as well make the best of it. If you don't want to mess with a garden, become

FIGURE 6.5. Over the years, our garden beds have become mostly composted rabbit manure. The only downside is that vigilance is needed to keep weed seeds from being carried in there. Weeds appreciate the rabbit manure compost's properties as much as plants we choose to grow.

friends with someone who does, and you can probably work out a pretty awesome barter system.

Manure will need to be removed from your barn no matter what you ultimately do with it. Left in an enclosed building, the manure will begin to break down, raising humidity to unhealthy levels for your rabbits.

An Alternative Electrical Power System

The last two barns we constructed were not connected to the electrical system—partly because we felt like we had given the power company enough of our cash, partly because we wanted to be more sustainable, and partly because we simply didn't have much more ability to add anything else to the existing electrical system.

Setting up an alternative power system does obviously require an initial investment up front. Depending on your needs or your location, solar or wind might be an option.

If you choose to go with a wind system, make sure the system will operate at the minimum wind speeds for your climate and area so that you can generate enough power.

There are dozens of good books out there on solar-powered systems,[1] so this will just provide highlights and a quick overview of how we set up our system.

First, determine how many volts your barn requires. What equipment you plan to run will determine the best system for you. If you plan to breed year-round and want to run lights, that will require more volts than if you just want to run a fan for ventilation. Ventilation is key, so first decide how much ventilation you will need.

Solar panels are usually connected to an inverter that "translates" energy from the solar panels into the battery

FIGURE 6.6. Our solar panels are mounted on the south-facing roof of the barn. And with Kansas's frequent summer thunderstorms, our hail screen is a must!

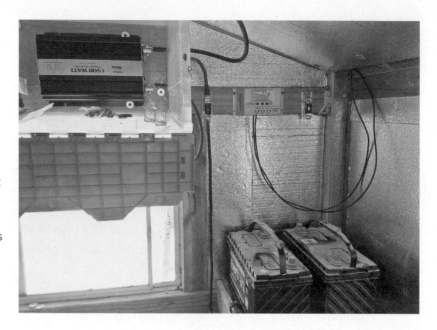

FIGURE 6.7. Below in the right-hand corner are two 12-volt deep-cycle marine batteries to store the solar charge. Directly above the batteries is the control box. In the upper right corner is the converter that enables fans and other 110V equipment to be plugged in. The converter is protected by a storage tote attached to the wall. This helps keep cobwebs and moisture from affecting it.

storage unit that will directly connect to whatever you are powering. If you don't have the inverter, you can't run plug-in (120V) items.

If an inverter isn't necessary, use your solar panels to charge batteries, and run 12V equipment directly from the batteries.

How much battery storage you need will depend on what you will actually be using in your barn.

Even without solar panels or an inverter, you can of course just run a few small things directly from a 12-volt deep-cycle marine battery. These batteries can be purchased locally, at any farm supply store. Small 12-volt fans, for example, can be hooked directly up to such a battery.

And, because Murphy's Law finds us all eventually, what is going to be your backup plan if the alternative power goes down, or your equipment fails for whatever reason?

Colony Raising

There is a lot of interest in raising rabbits in colony settings, or pasturing them for at least part of the time. The reasoning is that this allows rabbits to live a more "natural" life and exhibit "natural" behaviors.

While there are some definite pros to colony or pasture raising, "natural" life and behaviors also include exposure to disease, predation, and fighting.

Colony raising at its most basic involves rabbits that live in a group, eat from common feeders, have common nest boxes, and may or may not have the buck present with them all the time.

There are as many creative ways to build and arrange your colony as there are creative people to design them. Some clever people have repurposed old stalls or barns for their colony setting. Rubber mats on the floor can give the rabbits comfort without letting them dig...which they naturally will, with great determination. With outdoor colonies on the ground, you will need to take precautions to ensure that the rabbits cannot dig out of any enclosure. This can be done either by burying chicken wire, chain link, or some other durable wire around the perimeter of the colony. But it may be just a matter of time before the rabbits figure out a way around it. Or under it. I've read about outdoor colonies that laid chicken wire on top of the ground around the perimeter, but rabbits will figure out pretty quickly how to circumvent that.

Rabbits can also be much more territorial than people realize. It may take some time to match up a group of rabbits that will all get along. Rabbit fights can be vicious. When starting up a colony, spend some time making sure that all the individuals get along well, and that one rabbit isn't hoarding all the food and/or pistol-whipping the others. You may get lucky in the beginning and have a group that gets along, or it may take some time to find the right combination of individuals.

Digging burrows is one of the natural behaviors that people want to allow their rabbits to express, but be warned that may also include burrowing where you can't reach them. Raising rabbits in colonies *can* reduce the labor involved in raising rabbits. A well constructed colony with a group of rabbits that get along well and have adapted to the situation can be low maintenance, and infrastructure costs can be much less.

However, raising rabbits in colonies is very difficult on any sort of scale. First, the space required to allow the rabbits enough space to avoid conflict will be substantial. It will be difficult, especially if your setup allows them to dig their own burrows, to track the number of kits born to each doe, monitor their health, and check for kit deaths. The rabbits will either survive or not, and you may never know without excavating burrows. Monitoring litters can be much easier in a colony constructed in a building that does not allow digging, but your litters are still much more vulnerable.

Disease exposure is also a much more real risk in a colony setting. If one rabbit gets, say, a full-blown case of snuffles, the whole group is exposed, including all the litters.

You will also still have to clean manure from your colony. In an indoor barn, the manure and bedding will need to be taken out on a regular basis, and fresh bedding applied. Outdoor colonies will need some cleaning as well, although the elements will take care of some of the manure.

And, if everything works well in your colony, you may be overrun with rabbits. While this may seem on the surface to be a good thing, it can lead to problems as well. You won't be able to easily split up the litters by gender, and young, early-maturing bucks may be able to have their way with all the other females, leading to an even bigger population boom. This unmonitored breeding can take a toll on the does. Increased crowding with lots of rabbits can lead to fighting, and the risk for disease transmission increases exponentially due to both constant re-exposure and the stress of overcrowding. Stress lowers the immune system, and makes the animals more vulnerable to infection. Keeping adequate food and water can also be a challenge with an ever-expanding group of rabbits.

Pasture Raising

"Grass-fed" is both a good thing, and a consumer buzzword. Raising rabbits entirely on pasture sounds like a great thing, but it can be fraught with problems. Any time rabbits are exposed to the ground, they run the risk of picking up *coccidia*, a protozoan that can cause problems from gastric enteritis to liver damage. Enteritis often leads to the death of the rabbit; liver damage can result in the carcass being condemned upon processing.

Depending on the microclimate where you live, available forage, and stocking density, it may be difficult for your rabbits to get all the nutrition they need from a pasture. Even moving rabbits to fresh pasture daily doesn't always suffice. For the most part, you will need to plan on feeding some supplemental pellets, unless growing the rabbits out in a reasonable amount of time isn't a goal.

And, with drought being an ever more frequent concern in some areas, pasture may or may not even be an option.

Pens

We have always used pens for our rabbits, except during a grant study we did several years ago. It would have been impossible to produce rabbits on any scale with any other system.

One rule of thumb: You will always need more pens than you think you will. In fact, that should probably have been the title for this book! Especially if you intend to keep breeding stock, either as your own replacements or to sell, pen space fills up very quickly.

Rabbit Math
One buck = one pen
One doe = one pen
Each litter = four pens

No, my math hasn't gone wonky. Each litter can require up to four pens. Here's the logic behind the math:

1. When the litter is weaned, you will need to check the gender of the kits and separate them accordingly.

   There are multiple reasons for this. If the litter is large (say eight kits), then they will compete at the feeder and some will inevitably get pushed aside and not be able to compete well. Your goal in raising meat rabbits is to get as many pounds of meat in the most cost-effective amount of time, and this is best accomplished by getting a uniform batch of fryers to harvest weight at the same time.

Fast-growing rabbits will reach puberty at about 75 percent of their mature weight. For breeds like the American Chinchilla, this means that they can begin cycling just at the time they are hitting harvest weight. You don't want to find out that you've just harvested a pregnant doe.

2. If you retain breeding stock, they cannot be accurately evaluated until they are about 12 weeks of age, an age at which big, fast-growing rabbits can reach puberty. A promising doe can be tanked by being bred too early. While she can likely take care of the litter even if she has not reached her growth potential, she may not mature as well as she might otherwise.

3. Time always speeds up. Even if you think you're going to "get it done soon," soon often turns into too late. I don't know about at your house, but here there always seems to be some sort of time warp, where according to the list of things that still need doing, it should be early May, but according to the calendar, it's November. Take the time in the beginning to get that litter separated. They will grow better and more evenly, allowing you to accurately identify potential breeding stock without wondering if they would have matured better if they had less competition. And conversely, with no competition, you can see if that promising rabbit actually meets the standards you set, or stalls out at a certain point.

So, that's why in rabbit math, $1 + 1 = 6$.

Pen Wire

The only truly suitable wire for pen floors is ½ × 1-inch mesh (1.25 × 2.5 cm). Larger openings will be too hard on the rabbit's hocks by concentrating the pressure of their weight so that it's difficult for them to move comfortably. Smaller openings will not allow manure to pass through, resulting in a nasty buildup in the pen.

Hardware cloth or hail screen is often quite a bit cheaper than good wire, but it is far too rough and sharp to use as a floor. The rough surface of the wire will abrade the pads of the rabbits' feet quickly, and smaller gauge will eventually result in sore hocks. Sore hocks are incredibly painful for the rabbit, and can result in infection or damage to the foot that will likely require culling the rabbit. Smaller openings will also not allow waste to pass through easily.

One-half by 1-inch wire is definitely worth the investment. After all, your rabbits will be spending a great deal of time in that pen, you want them to be comfortable—not only because it's the right thing to do, but also a happy comfortable rabbit will grow better, and perform better.

The sides of the pen can be the larger and less expensive 1 × 2-inch (2.5 × 5 cm) wire, however, as can the top. One caution, especially for doe pens, is that the sides should have some sort of barrier to prevent newborn, escaped kits from getting into an adjacent pen. It is amazing how far a little kit can squirm trying to make its way to someplace other than where it is.

Hardware cloth is also unsuitable for pen sides, as it will not stand up to rabbits messing with it on a regular basis. Once they figure out they can pull the flimsy wire apart, they will tear it up in no time.

About the only thing we can recommend hardware cloth for in the rabbitry is covering wood to keep rabbits from chewing on it, or its intended use as hail screen or window mesh.

Pen Dimensions

Pens work best longer than they are deeper. Six inches does not sound like a lot, but when the rabbit retreats to the back of the pen to avoid being grabbed, it's a frustrating distance.

Our pens are 24 inches deep, 33 to 36 inches wide (for does, litters and nest boxes), and 24 inches tall (61 cm deep, 84–91 cm wide, and 61 cm tall).

You can buy pens already made, but they are usually only 18 inches (46 cm) high. This does not allow the rabbits to fully stand up on their hind legs, which they enjoy doing to look around.

One-half by 1-inch cage floor wire usually comes in 30-inch-high (76 cm) rolls. To cut down the size, it is necessary to cut off 6 inches (15 cm) from the roll. This wire can be recycled as baby saver wire, placed around the bottom of the cage sides. You can skip this step, but if you are short or have short arms, be aware it's going to be challenging to grab a reluctant rabbit.

Buck pens can be slightly smaller, since they will not have a nest box to manage. We have a separate buck barn that houses all our herd bucks. Those pens are 24 inches deep by 24 inches high, and 30 inches wide (61 cm deep, 61 cm high, 76 cm wide).

If you will only have one building it might be much easier to make the pens all the same size, so you have more options for moving animals around.

When cutting the door into the pen, make it large enough to be easy to get rabbits in and out, and also to get the larger nest box in and out. About 15 inches (38 cm) square should be adequate.

When putting the cage together, use cage rings rather than flat clips. Flat clips will rust easier, and can be more challenging to replace. Rings are a bit more expensive, but they will last longer and be easier to use.

Depending on what sort of feeder you use, it will be necessary to cut a hole in the cage to install it.

Outside mount feeders are usually the best bet. Crocks can be spilled, pooped in, and need frequent cleaning. They also take up valuable floor space in the pen.

When mounting a feeder on the cage front, install it in the doe pen about 4 inches (10 cm) off the floor. You may be tempted to put it down low for the kits to access, but it is best for them to have to reach up into the feeder to eat. This will help keep them from crawling into it.

Constructing Pens

One of the simplest and easiest ways to create a run of pens is to lay out a single length of pen wire the length you want your run to be, and then use three more lengths of wire to create the top, back and sides. Dividers can then be placed in the desired location (usually before the tops go on) and doors cut into the front.

Shaping the pen roof like a Quonset™ hut can make the most efficient use of wire, and if two rows are placed back to back, the center can be used as a hay feeder. But the sloped back will make it difficult to have

FIGURE 6.8. This stretch of pens is a single run of 1 × ½-inch wire suspended between T-posts, and anchored to the wall for stability. Two more runs of wire form the roof and back. A couple of support wires run from the roof of the pens to the roof beams. Dividers are individually cut to fit and installed where needed. While it looks like it should fall down, this stretch of pens is surprisingly strong. Having fewer support posts makes cleaning underneath simple and easy.

FIGURE 6.9. Another barn with similar pen setup. The top of these pens is shaped like a Quonset™ hut. This saves wire, but reduces the useable space of the pen. The middle between the two pens can be used as a hay feeder. In this photo, it is empty between processings.

room for a nest box. If you have a row of dedicated pens for feeding out fryers, this design can work quite well. Again, though, accessing the center row for cleaning can be difficult for shorter folks.

In general, single layer cages will be better for the animals than double or triple layers. Eric always says that being the bottom rabbit in a stack of cages must be like living in the basement of an outhouse. There is a reduced amount of humidity in a single layer, and pens are also easier for the farmer to access. A double layer of cages will require reaching up to get into the top row, and bending over to get into the bottom row. It may make it a little more challenging to check and observe nest boxes as well.

Snuffles is often a problem with multiple tier cages; this is likely due to the increased numbers of animals in a smaller space. If you choose to go with a multiple tier cage system, be even more vigilant about ventilation and air quality in the barn.

Outdoor Pens

Pens can also be constructed out of doors to take advantage of natural shade and airflow in the summer. In an area where the winters are mild, simple windbreaks may be enough to keep the rabbits comfortable all year long.

In much colder climates, it's a balancing act between shielding the rabbits from cold wind and precipitation, and making sure the airflow is adequate to avoid building up ammonia fumes from urine. This will take some management and due diligence on the part of the farmer to monitor and make sure the situation is good for the rabbits. Remember, if you step into the rabbit area and can smell a strong odor, the rabbits have been breathing that all along. Some odor with animals is unavoidable, but it shouldn't make your eyes water on a regular basis.

It is not necessary to take your rabbits inside when it gets cold! They handle cold weather much more easily than hot, and with good shelter and available food and water, they can handle the cold just fine. Taking them in and out exposes them to dry household air, and forces them to readjust to the different environmental conditions every time they go in or out. If you live in an area where the cold is a major concern, you might be better off constructing a simple barn for your rabbits rather than keeping them in outside pens.

Predation is also a concern with rabbits in outside pens. Rabbits are a tasty treat for many carnivores, and wire pens will not stand up to determined dogs or coyotes. Discussing dead rabbits and guilty dogs is not the way you want to meet and interact with your neighbors, although I gave a couple of our less responsible neighbors a few good Crazy Rabbit Lady stories back in the early days of our livestock ventures. And they did learn to keep their dogs at home.

Insulation

Whether you build a new barn or retrofit an existing structure, take the time to insulate the building. With all the modern types of insulation available, installing it is relatively inexpensive, and easy to do. One person can handle even a 4 × 8-foot piece of polystyrene foam insulation sheeting without too much difficulty, and two people can make quick work of installing several sheets.

Insulation will not only keep the barn warmer in the winter, it can help keep it cooler in the summer. A metal roof on a building can increase the temperature inside a significant amount, and even with good ventilation, moving air that is warmer than necessary defeats the purpose of fans for cooling.

FIGURE 6.10. Insulation going up in one of the barns. This was an upgrade from Masonite sheeting we got for cheap early on in the evolution of our barn design and learning curve. While the cost outlay for the masonite was minimal, and it did a fine job of insulating the roof, it soon began to sag and absorb condensation, ultimately requiring replacement.

We've tried several versions and types of insulation over the evolution of our rabbitry, and the best value for us in terms of cost, insulating capability, and ease of handling has been the ½-inch (1.25 cm) Styrofoam™ sheeting. Its metallic covering has the added benefit of brightening up the barn. The drawback is mice also like it, but if you keep up with pest control, the impact is minimal.

CHAPTER 7

Basic Rabbitry Equipment

This chapter describes the basic equipment we use in our rabbitry every day. There are, of course, all sorts of things you can purchase to use with your rabbits, but having these basics should get you off to a good start.

Scale

A good scale is one of the best investments you can make in your animal's health. It provides objective data on whether or not your rabbit is too fat or underweight, how closely it adheres to a breed standard, and many other criteria. Weighing can help you know when your meat rabbits are ready to process, so that valuable time is not wasted processing animals that are not quite ready or continuing to feed animals past their optimum processing weight.

For a small rabbitry handling only a few animals a year, a bathroom scale can work quite well. If you will be processing on a larger scale, a platform scale with separate digital readout saves frustration. The nice thing about a digital scale is that the reading unit can be mounted on a wall—where it is much easier to read and out of the way of rabbit urine and feces.

Either way, you will need a basket or crate to put the rabbit in when you weigh it. Otherwise, the animals' inevitable movements will not let the scale read accurately. Rabbits may be a little restless at first, but will get used to being handled and weighed, taking the process in stride.

FIGURE 7.1. This spring scale was purchased at a local outdoor goods store. A wire basket attached to the platform, and this will provide years of use.

FIGURE 7.2. Digital scales are preferred for weighing fryers, as the digital readout has a lock function so you can get an accurate weight.

If you do get a platform scale, it might be wise to put some non-skid material on it, as most scale platforms with shiny stainless surfaces can be slippery, and cause the rabbits to scramble to keep their footing, even causing the basket to move around.

Another inexpensive alternative is a platform nursery scale. With a piece of carpet attached to the pan, this will give a non-slip surface, and work well for weighing rabbits of all sizes.

Frequency of Weighing

Weighing animals regularly is a step many people are tempted to skip, but accurate weights and regular recording of weights can help any rabbit program to improve.

Take each rabbit's weight at weaning, and then again closer to processing. Rabbits that outgain their littermates can then be given a chance as breeding stock, and rabbits doing poorly can be culled.

FIGURE 7.3. These scales are relatively inexpensive and never need batteries!

Weighing a doe at breeding and weaning can ensure that her condition does not get pulled down too much, and that she can carry her next litter successfully.

And bucks can be monitored for health as well as making sure they don't get to big or heavy for their breeding job.

Nest Boxes

As a general rule, we feel that there are no absolutes, and what works for one rabbitry might not work for another. Typically, there are dozens of ways to do things and dozens of types of equipment that work—with one exception.

We strongly and passionately differ from the conventional wisdom in what to use for a rabbitry nest box. We are definitely not fans of the metal nest boxes that are the standard choice—or in-floor, drop-down nest boxes for the same reasons. Here's why:

► The standard nest box available from a farm supply store is not big enough for large meat breed rabbits and their litters. A doe jumping in and jumping out can injure her litter; she can step on them because she just can't avoid them.

► Metal boxes get too cold in the winter and hot in the summer.

► Wood floors trap dirt, urine and bacteria, and because wood is porous, these boxes are not easily disinfected.

► Nest boxes with wooden sides are also too hard to clean.

► Open-topped boxes don't provide doe rabbits the mental security and privacy that, as a prey animal, they are hardwired to want.

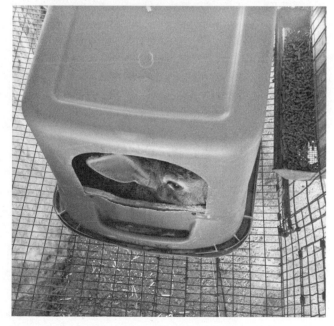

FIGURE 7.4. All this American Chinchilla doe needs now is a "Do Not Disturb" sign!

FIGURE 7.5. A summer nest box in which ventilation holes keep air moving, while still giving this Silver Fox doe her privacy.

▸ We have had excellent results with homemade nest boxes created from 10-gallon (40 L) gray plastic storage boxes we purchased at our local discount store. These plastic totes are relatively inexpensive, last for several years, and can be cleaned and easily disinfected between litters.

Since we have litters year-round, we have designed a couple of styles with the seasons in mind. The winter nest box is enclosed to retain heat, while the summer box has 1-inch (2.5 cm) holes drilled on the sides with a paddle bit to allow for ventilation. The nest box bottoms (which are the lids of the totes) have small holes drilled in them to allow drainage. The nest box top (which is the bottom of the tote) gives the doe a resting place, and an area to escape the kits if she feels the need.

In the initial months of our rabbit venture, we used the metal nest boxes (simply because we had inherited some), and we had poor luck with does not wanting to use the nest boxes and having their litters on the wire. It got pretty discouraging. Eric even got up at all hours of the night to go out and check on the does, and still regularly found them with dead litters.

There had to be a better way.

Remembering that the rabbit nest boxes his grandfather used were old wooden apple crates, Eric started looking for a modern alternative and ultimately settled on plastic storage totes.

After some design and rotary power tool work, he had a couple ready to go. When the next litters came around, we waited, fingers crossed. When Eric went out to check at 2 a.m., he was amazed that not only had the does used the darn things, the babies were snuggled down nice and toasty in perfectly built nests.

We haven't used anything else since. Occasionally a doe will not want to use the box, but this is rare; those who refuse are usually does with poor maternal abilities. We estimate our death loss due to kits being born on the wire is around 1 percent or less.

The down side of plastic totes is that some chewing will occur, but this has not been a significant problem. Fastening a short piece of a 1-inch (2.5 cm) furring strip (not treated wood) along the bottom of the front opening can give the doe something healthier than plastic to chew on. Eric calls this little piece of wood the doe's "worry stick." It's easily replaced once it has been chewed too much.

I know, lots and lots of litters are born every year in the conventional nest boxes. But ours have solved a lot of problems and concerns for us, and we have no desire to go back to conventional ones.

Nest Box Bedding

When creating a warm and secure nest, the doe will pull fur from her chest and underarms, and mix it with whatever material you provide in her box. There are several choices for bedding material to give her, and in our opinion some of them work better than others.

Hay

Clean grass hay can be a good choice for bedding. Rabbits can munch on it if they choose, and the extra fiber can be good for their digestive system. Watch, however, that they are not eating their bedding at the expense of their regular diet.

Grass hay can become damp and soiled easily, and is not the most absorbent choice for bedding. If you use hay, be sure to change it out frequently; pull any damp clumps out. A wet, messy nest box can be unpleasant at best for the doe, and at worst can harbor bacteria and even possibly allow flies and maggots to breed there. Infested bedding is not the environment you want for rabbit kits.

Another caution about grass hay is that it may contain toxic grass and weed seeds. This is not a problem for the rabbits, but if you add your bedding to your compost pile or garden, weeds may come along for the ride.

Straw

Straw can also be used, however it is not absorbent enough to really keep the nest box dry. Straw should also be very clean, as it tends to hold dust and dirt well. It also can become damp easily and require frequent changes. Straw can also generally come with seeds eager to find their way into your garden. And once damp, straw loses its insulating capacity quickly. Straw is also a hollow tube, perfect for harboring bacteria and moisture so well that it cannot escape easily.

Make sure the hay and straw your use are clean, dust free, and mold free to avoid eye irritants, or a doe ingesting something that is not good for her.

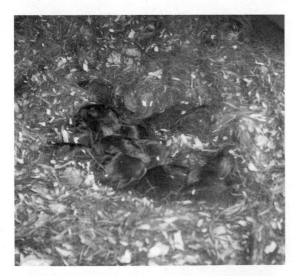

FIGURE 7.6. Shavings cling to the kits.

Shavings and Sawdust

Shavings are commonly used as bedding, however we do not like them at all for nest boxes. They are usually dusty, and in the case of sawdust, can irritate eyes and cause respiratory problems. Shavings are highly absorbent, but not worth the risk of the irritation for rabbit kits. Sawdust will also become matted easily, and require frequent changes.

If shavings are all that is available, make sure they are as clean as possible, and take care to monitor the nest box and the kits frequently.

And at all costs, avoid cedar shavings. Cedar contains a compound which can serve as an anticoagulant/irritant to navel stumps, and can even cause death for the kits.

Shredded Newspaper or Household Paper

Newspaper is not recommended at all for nest box material. The inks can be toxic, and the paper itself is often treated with chemicals. Do not use household paper either, these inks can also be toxic and the paper highly processed and not as absorbent.

Shredded Feed Sacks

An alternative to purchasing shavings is to shred paper feed sacks and use them for bedding. This works well as the paper has no seeds which can transfer into the garden. Paper sacks are usually printed with soy ink, and the paper has been minimally processed. It composts quickly.

The material is highly absorbent, and has little dust to irritate eyes or respiratory systems. It mixes well with the doe's fur and provides natural insulation. And after the initial outlay for a mechanical shredder, it costs very little to make a large and continuous supply of bedding.

Recycling feed sacks this way also keeps the sacks from getting thrown out, and you don't have to find some other way to dispose of them.

FIGURE 7.7. Shredded feed sacks are cleaner, and even with the time necessary to shred them, a cost saving.

FIGURE 7.8. Shredded feed sack mulch in the garden.

To shred your feed sacks, cut them into 6-inch (15 cm) strips and feed these through the shredder. A heavy-duty shredder can handle one section of bag at a time. We use a paper cutter to cut the bags since we use so much bedding, but for smaller operations heavy-duty scissors can do the trick.

Even a heavy-duty shredder will get hot, so give it a time-out occasionally if you are shredding a bunch of sacks at one time. Blowing the inevitable feed dust out on a regular basis will add to the life of your shredder, too.

It's best if you can avoid using feed sacks of other types. Depending on the feed there may be different additives that you may not be aware of. Avoid feed sacks that have plastic linings. And do make sure you use feed sacks that have not been stored where cats, rats or other predators can crawl all over them, leaving their scent and other potential bacteria or disease.

Another distinct advantage to the shredded feed sack bedding is that when it invariably makes it into the compost, we do not have to worry about seeds from straw or hay making into the mix. Since one of the other products we market is rabbit manure compost, our customers appreciate this. The bedding also works as a biodegradable mulch around the garden.

Feeders

Feed Crocks

Crocks have been a universal staple in the rabbitry for decades, and they still can serve a function. They allow rabbits to eat around the whole dish; they should all be able to access feed.

But there are a couple of caveats. Kits can drag feed out of the dish, which will fall through the wire and be wasted. Kits can also climb into the dish and think it's a pretty cool litter box. And feed in an open dish is more likely to become damp and spoil.

Crocks also can take up valuable floor space in the pen, and as the kits get larger and more boisterous, a crock can easily get upended.

Cage Mount Feeders

Metal feeders can be mounted on the wall of a pen, thereby not taking up any floor space at all. Most metal feeders will have a mesh bottom, which will allow dust and fine particles to fall through. Kits cannot climb in wall-mounted feeders as easily, although there is always one Houdini in the bunch every now and then.

There are several brands of metal feeders, and all of them are pretty much the same. And eventually the mesh can wear out in the corners, requiring the feeder to be replaced, or a crafty repair job to patch the hole.

A 3½-inch (9 cm) feeder will work for a single rabbit, but for multiple rabbits, an 11½-inch (29 cm) size will be much better.

A *creep feeder* can also be a handy investment, this can give larger litters a chance to supplement their diet and take some of the pressure off mom. We have also been able to use it in the event of a doe dying before her litter is quite old enough to wean. Giving the young kits a small amount of Calf-Manna can help them get over the stress of early weaning and go on to thrive.

FIGURE 7.9. This 11½-inch feeder installs easily in an opening cut in the front of the pen. It allows loading feed from the outside, keeps feed clean, and allows plenty of room for everyone to get up to the feeder and eat.

FIGURE 7.10. Kits at a creep feeder.

FIGURE 7.11. Whenever possible we use drinker bottles. The lids keep flies out, and the water stays cleaner.

FIGURE 7.12. Cup drinkers.

Watering Equipment

Crocks can also been used as water bowls for rabbits, but there are a cautions about using them for water.

Rabbits will regularly stomp in their water with dirty feet, contaminating it and contributing to the spread of coccidiosis,[1] and just generally making the water unpleasant to drink.

Bowls of water can of course also be spilled much easier, and it's difficult if not impossible to keep flies and dirt out of them. Rabbits have also been known to use their water crocks for litter boxes.

Water bowls should be rinsed daily, and disinfected regularly.

Water Bottles

A better choice is a drinker bottle that can be mounted on the outside of the cage.

These bottles can be adjusted to the height of the kits easily. Kits that start drinking water earlier will start eating solid food earlier, gain weight better and take pressure off the doe.

Flies cannot get in a drinker bottle as easily, and the water stays clean and fresh longer. The bottles also do not take up space on the pen floor.

Water jars do not work as well in subzero temperatures. You can switch out to cups during cold months, or keep the barn temperature above freezing. If you have bitterly cold winters, this may not be cost-effective, but in more temperate zones a heat lamp on a thermo cube may be enough to keep the water clear.

If water keeps freezing, be sure to offer the rabbits water in the morning and in the evening to make sure they get ample opportunity to drink. Ice is not a substitute for water. No animal can get enough water from licking ice to keep their systems functioning optimally.

Automatic Watering Systems

The importance of maintaining fresh, clean drinking water can not be overstated. If rabbits don't drink, they won't eat.

An automatic watering system can be easily and inexpensively constructed. These systems involve a container of water that can be suspended above the rabbit pens, and a system of tubing or lines to carry the water to the individual pens.

Most of the tubing and fittings can be obtained a poultry supply houses. Drinker nipples are available from the larger rabbit supply companies.

Be sure to install the tank high enough that the lowest level of the tank is at least 1 to 2 feet (30–60 cm) above the pens.

This of course presents a challenge if you are short like I am. Eric can easily reach the top of the tank to fill it; me, not so much. I either have to have a five-gallon (18.9 L) bucket handy for a step stool or simply wait for him to do it.

FIGURE 7.13. Automatic watering setup.

FIGURE 7.14. A supply line for water lines to individual pens is shown near the top of this photo.

Whatever tank you use should be opaque, and dark enough that sunlight does not encourage algae to grow. Keep the lid on the tank to prevent dirt, insects, and other debris from getting into the water. And, once or twice a year, take the tank down, clean it, and disinfect it, either with bleach or good old sunlight.

With a little practice, you will figure out about how long it takes for the rabbits to empty the tank. If you time refilling for when it's had a chance to empty, the tank will keep the water fresher.

These tanks are great for the barns that have multiple rabbits in each pen. The water won't run out before the least assertive rabbit has a chance to drink, and in the summer, this makes sure there is a plentiful supply of water for all.

Unfortunately winter and freezing temperatures limit year-round use for us, but the tank and line system makes life much simpler in the summer. And really, it's only two or three months out of the year we can't get use out of this system.

Do watch out for leaking drinker nipples, as they will waste water, and also add moisture to the ground—which increases humidity in the barn. And also gives flies a prime breeding ground.

We use an automatic watering system for our group pens, and keep cups or jars on the breeding stock all year-round. This allows us to monitor how much the animals are drinking. A rabbit that is off water either already has some health issue, or soon will. The jars also allow us to make sure the kits are drinking too.

In the winter, when it is too difficult to use a hose and keep it from freezing, we use a wheeled water cart that holds eight gallons (30 L), and keeps the chore from becoming overwhelming. A garden watering can with the diffuser removed can work for smaller rabbitries.

Fans

Ventilation is one of the most critical factors in a successful rabbitry, and one of the most often overlooked.

Rabbits cool themselves not by sweating or wallowing, but by radiating heat out of the veins in their ears.

Rabbits also cool themselves, but by a lesser degree, by panting. This type of passive cooling exchanges body heat via exhaling the warm air from the lungs, and breathing in cooler air.

For cooling by breathing to be effective, there needs to be consistent and constant air movement around the rabbits that pulls heat away.

In the absence of electricity in your barn, a prevailing breeze can work just as well, but Murphy's Law dictates there will be days when you need that breeze, and it fails you. Make sure you have a backup plan, especially if you live in a climate where the summers tend to be hot.

While fans provide an effective boost to natural airflow, be careful when selecting a fan. Cheap box fans without enclosed motors will attract a dangerous amount of fur, dust, and cobwebs; more than one barn fire has been started by an overheated fan motor. And at least you will want to avoid replacing fan after fan.

FIGURE 7.15. Winter watering equipment: eight-gallon watering cart plus garden watering cans.

FIGURE 7.16. This photo clearly shows veins in the rabbit's ear.

FIGURE 7.17. This fan with fur buildup is a fire looking for a place to happen.

FIGURE 7.18. This fan is installed slightly above floor level. It draws ammonia-laden air out without creating too much of a draft.

Spend extra effort and money to purchase a fan with an enclosed motor. The extra money spent will pay off in better operation and less worry.

When installing the fan, make sure to put it where it does not blow directly down on the rabbits, but rather moves the air over them. This will help pull the heat away from them and allow them to cool down.

Installing timers for the fans can help cut down on the cost, and stop them from running all the time in the winter months. Yep, ventilation is even more critically important in the winter. Timers also can be set to reduce the amount of time the fans run at night, when they are needed less. Select a timer that allows you to set intervals of 15 minutes or so. This will allow the movement of enough air to keep the quality good in the winter, without moving too much cold air through the barn.

Misting Systems

In hot climates, misting systems can be invaluable to help cool the rabbit barn. These systems produce a fine mist, that as it evaporates, pulls heat out of the air.

You can use the same misting kits that are used around a deck or pool in the summer to help cool people off. These kits can be found online, or at retail stores that carry seasonal items.

Kansas gets some extremely hot weather during the summer months. The misters set up along the outside of the barns have helped keep the rabbits comfortable and productive in the heat.

We set up our system on the outside of the barn rather than the inside.

FIGURE 7.19. Misters are hung above the windows on the outside of our barns, so that cooler air gets pulled inside.

This is not to say a mister cannot be used inside the barn, but any mister nozzle must be attached to a high-velocity fan to move moisture through the air to cool it. Without a fan, mist will simply create damp, humid unpleasant conditions inside the barn and increase humidity too much. You will completely undo any benefit you are trying to achieve.

The mist needs to be a fine mist, not the type that comes from a spray nozzle. The idea is to not to get the rabbits wet, but to cool the air temperature, via evaporation, on the way into the barn.

We use misters when the temperatures reach 85°F (29°C) or higher. They get turned off at night, as the temperature drops.

However, misters are ineffective when the humidity is high. At that point it is critical to keep fans running and air moving over the

FIGURE 7.20. This mist nozzle is attached to high-velocity fan.

rabbits. We don't use ice or frozen water bottles, we simply have too many rabbits for that to be effective at all. Ventilation is everything.

As you can imagine, misting in summer is hard on the water bill for the farm. We are on rural water, since the solid bedrock 3 feet (0.9 m) under the topsoil on our property makes it impossible to dig a well. But even though water usage doubles in the summer, it only takes saving four rabbits to justify that water cost. That, and it is senseless to let rabbits suffer from the heat if a mist system can help at all. In order to recoup some of the water and at least feel a little greener, we've been doing some container gardening, and placing pots with zucchini and other vegetables under the windows below the mist nozzle heads. Container plants dry out so fast they are hardly worth it in the July heat, but keeping them under the mist system helps immensely.

Tattoo Equipment

At some point, it may be necessary for you to tattoo the offspring or breeding stock that you retain permanently. Keeping good records is important for improvement in your rabbits, and while it may seem like with a few rabbits it's possible to know who's who, tracking the generations can get overwhelming pretty quickly.

A good set of *tattoo pliers* is relatively inexpensive (less than US$60), and they will generally come with a set of digits. Including a complete set of letters adds a bit more to the cost, but if you come up with a simple numeric system you won't need letters, at least not for some time. These pliers are available online, but if you attend any of larger rabbit shows such as the ARBA nationals, there are usually several vendors there, and you might be able to get a better deal.

FIGURE 7.21. This photo shows both types of tattooer. The toothbrush is for working ink into perforations in the ear after the plier is applied.

FIGURE 7.22. Rabbit in tattoo box.

The good news is that, with proper care, pliers and letters will last a very long time. I have literally tattooed hundreds if not thousands of rabbits over the years, and the pliers are still as sharp as ever.

Some folks swear by pen tattooers that are on the market. I tried one in the early days, and for me it was slower and not as effective. Many rabbit folks claim the pen doesn't hurt as much, but I had a few drama queen rabbits that really hated the pen, and it was hard to get a good tattoo. Plus, my writing isn't exactly calligraphy quality, so we switched to the pliers. I will occasionally have a rabbit scream when I apply the pliers—and they will occasionally hit a blood vessel—but the pain appears to be over quickly, and the rabbits are back to eating as soon as they are placed back in their pen.

A *tattoo box* is strongly recommended if you are going to tattoo many rabbits. We have the one that used to belong to Eric's grandfather. Not only is a sentimental piece of equipment, it's still performing admirably after 50–60 years of use.

It's possible to wrap rabbits up in a towel or sweatshirt sleeve to tattoo, but unless you wrap the rabbit securely and squarely, it's possible to have them kick and twist hard enough to break their back…which is simply horrible. Trust me on that one. Since we switched back to using a tattoo box, we have not had a single rabbit injured during tattooing.

CHAPTER 8

Diet and Nutrition

What do you plan to feed to your rabbits? Unless you have experience in nutrition, ration formulation or just enjoy doing it the hard way, it's probably better to start out with a commercial pellet.

We use a pelleted diet for a variety of reasons:

1. It is formulated and balanced to the rabbits needs.
2. The contents are evenly mixed in each pellet so the rabbit gets a complete diet with every bite.
3. With several hundred rabbits at any given moment, our time is a better spent on other tasks than mixing feed.

Yes, it's possible to feed rabbits on simply what you grow in the garden or harvest in the pasture; the satisfaction of having a product raised completely without outside inputs can be enormous. But care must be taken to assure a complete diet for the health of the animals. Mixing feed is also time-consuming, but if time is not an issue for you, go for it.

Just make sure to do your homework. Don't assume because wild rabbits run around eating grass and clipping daisies that domestic rabbits can get by on the same thing. With the rising cost of feed (which isn't likely to change any time in the near future), going to a home-grown diet can be awfully appealing. Just be sure to weigh all factors carefully before jumping in.

The Rabbit's Digestive System

The rabbit's digestive system is rather unique among domestic animals.

Rabbits are not *monogastric* (with a single-chambered simple stomach like pigs, chickens or humans). They are true herbivores, like cattle

FIGURE 8.1. Pelleted feed.

or sheep, but do not have those animals' large rumen or multi-chambered stomach to ferment and digest cellulose.

As nonruminant herbivores, rabbits have developed an enlarged hind gut and a unique digestive strategy to make use of high-fiber materials and plant cellulose.

The *cecum* in rabbits serves a similar function as a rumen, and performs microbial fermentation. But because of the smaller size of the animal, it cannot do the complete job of digestion.

Rabbits have developed the behavior of *cophropagy* (eating their own feces) to solve that problem.

Coprophagy

In most species, cophrophagy is a negative behavior. Its generally seen at best as a vice, a bad behavior brought on by boredom, or at worst, a desperate attempt to get nutrients that it is not getting in its diet.

For the rabbit, neither is the case. Coprophagy for the rabbit is a behavior developed in order to fully digest its food and get the maximum amount of nutrition out of its diet.

Rabbits have a relatively short, simple digestive system. This means that without cophrophagy, they would be unable to fully digest their food, or have to spend more time eating, a strategy that would render them more vulnerable to predators.

As the rabbit eats and food enters the digestive system, it passes first into the stomach. There, acid is secreted that helps begin breaking down the complex fiber that should be the bulk of the rabbits diet. From there, it enters the small intestine. In the small intestine, the more easily absorbed nutrients are taken up, and the rest of the feed enters the cecum.

The cecum functions in much the same way as a rumen does in a sheep or cow. It is host to a myriad of species and types of bacteria, and those bacteria then do the work of breaking down the feedstuff further.

But a rabbit's cecum is much smaller, and has greatly reduced capacity for volume, than a proportionally sized rumen.

Activity in the cecum creates either the soft *cecotropes*, or what becomes the harder, more familiar actual manure pellets.

Both then pass through the large intestine.

These pellets, called *cecotropes*, are then directly consumed straight from the rabbit's anus, and enter the digestive system again for further breakdown.

No one is exactly sure how the rabbit knows which pellets to consume and which ones to let go. Circadian rhythm certainly seems to play a part. In studies that tracked when rabbits eat and then within what number of hours they begin coprophagy, rabbits seem to consume the majority of the cecotropes during the daylight hours. But even rabbits kept on a 24-hour light cycle still keep to the same rough schedule. Smell doesn't seem to play a part in choosing, either, so it might be some combination of feeding schedule and the sensations in the intestines.[1] For now, the timing of accurate coprophagy is a secret the rabbit keeps for himself.

Pelleted Feed

We feed our rabbits a pelleted feed that is made at a co-op in our area. This feed as closely approximates a historic diet we found in one of the dozens of historic rabbit books on our shelves,[2] and met the criteria we wanted for our herd.

It's impossible on the scale of production we have to not use a pelleted feed. We currently go through a ton of feed per week to keep the number of rabbits we have fed and happy. Harvesting that much food by hand would be impractical to say the least.

In the early phases of our operation, we did try feeding a grain mix; mixing a certain proportion of whole grains in a large feed grinder and hauling it into the barns.

We abandoned this approach as our operation expanded.

Rabbits are also extraordinarily picky eaters. They have around 17,000 taste buds in that little mouth, and they know what they prefer.[2] When we mixed grain feed, the rabbits picked through the grain, finding the pieces they liked best, and dropped the rest onto the ground. They weren't getting all the nutrients they needed, and their growth was hit or miss.

We also had some lovely wheatgrass crops growing in our rabbit barns.

So, while loading and unloading that pellet feed weekly is laborious, in the long run it saves time, money, and provides the balanced diet that production rabbits need.

Alternatives to Pelleted Feed

There is however, a lot of interest in avoiding pelleted feeds and feeding rabbits natural, foods harvested at home. Since this is something we don't have hands-on experience in, and because availability of foods differs widely from place to place, we offer a broad overview of things to keep in mind if you choose to go this route.

High Fiber

Rabbits depend on high fiber for the bulk of their diet. The basis for their diet should be good, high-quality grass hay. This will be hay that is clean, free from dust and mold and of a sufficient protein level to sustain their systems.

Greens

Avoid greens such as iceberg lettuce. It is very high in water, and can cause loose stools and other digestive upsets. Dark greens such as kale and leaf lettuce are a much better choice. They are high in vitamin A and other nutrients. But be warned that once greens start to wilt, they can begin to ferment. Only feed what is fresh and what the rabbit will eat in about 15 minutes of time.

Treat Foods

Treats such as carrots, fruits and starches are high in sugar, and should only be fed in extremely limited quantities. The high sugar content can be extremely detrimental to rabbit gut health.

FIGURE 8.2. Rabbits eating fresh grass.

Rabbit Food Caveats

- ▸ What wild rabbits are able to eat and what you can provide for your domestic rabbits is not the same. The animals aren't the same species. Yes, they share certain similarities and can eat some of the same things, but they just don't. Also, as wild rabbits hop around and graze, they are able to eat the forage in its freshest state, not after cutting and sitting around for a time.
- ▸ Don't assume, because your rabbit is a herbivore, you can dump a box of leftover restaurant vegetable scraps in their pens for them to eat. At best, the animals won't get good nutrition. At worst, the scraps will sit on the floor of the pen and wilt, fermenting and drawing flies. Yes, we've seen it.
- ▸ Yard clippings are too fine, and by the time they get to your rabbit, will have started to wilt.

Other Forages and Herbs

There are lists and lists of foods such as herbs and other plants that rabbits can be fed. As we have no experience with these, we can't say what does or doesn't work. If you choose to go this route, make sure to introduce foods slowly. And, although it sounds brutal, you might want to do some experimentation with rabbits that aren't your best breeding stock, until you see how well they tolerate what you are feeding them.

Our Experiment with Rabbit Tractors

One of our early ventures involved trying to create some sort of *rabbit tractor* to see if there was any difference in the growth rate, nutritional value, and general well-being of rabbits on pasture compared to the pen system.

We knew we didn't want the rabbits directly on the ground for several reasons, a primary one being we didn't want the grass torn up by their digging.

We designed a system whereby we created frames of PVC pipe, attached ½ × 1-inch (1.25 × 2.5 cm) pen wire to them, placed them on the ground, and created a moveable shelter box to house them in, and placed it on a rail.

As far as the animals' comfort went, the moveable boxes worked well. The rabbits were able to have plenty of shelter from the elements. The shelters got moved along the track daily, so fresh grass was always available.

In the interest of scientific experimentation, we performed two trials using three different feed strategies: one just with the pasture grass, one with supplemental oats, and one with supplemental rabbit feed.

We weighed each rabbit weekly, and tracked weight gain.

As you might expect, the group that was supplemented with pellets showed the best weight gain. Second was the group with oats,

FIGURE 8.3. Rabbit tractor.

and lastly was the just grass group, far behind the others; it took about an extra three months for grass-fed animals to get to harvest weight.

There were some standout individuals in the trials, and if this was a method we intended to continue with, it would have been wise to retain those animals for breeding stock to see if there was some sort of genetic trait we could capture.

As it was, moving the tractors was labor-intensive, and as Eric was doing everything on the farm solo, continuing to use rabbit tractors wasn't something we could pursue.[3]

What To Do if a Rabbit Is Off Feed

The first temptation is to find it some treat or green food that it has liked in the past, just to get it eating. This can, however, make a problem worse, since the starches in those foods can irritate the digestive system and cause loose stools.

First try offering some good, clean grass hay, or some rolled oats. Both of these feeds are high-fiber (which will soothe an upset digestive tract) and are especially palatable (especially the rolled oats) to rabbits.

Also, be sure to check the water supply! Rabbits should have access to fresh, clean water. Check the bowl to make sure it isn't fouled, or check the nipple on the drinker bottle or water line to make sure they aren't plugged. It doesn't take long without water for a rabbit to become dehydrated and not want to eat.

If the rabbit has water, but still is off feed, check the consistency of its droppings on the floor. If they are runny, offering high-fiber feeds is even more important.

I have only on a couple of occasions (in decades!) seen Eric offer a small handful of grass to a rabbit that was off feed. These were not yard clippings, but rather tall leaves of high-fiber, long-stemmed brome grass that could be tied up into a knot and left in the pen. On these couple of occasions, if my memory serves, the fresh brome grass did help get the rabbit back on feed, but it took some time for the rabbit to recover fully.

CHAPTER 9

Hands-On Evaluation of the Rabbit

When dealing with purebred, pedigreed stock, hands-on evaluation of the rabbit will always begin with the Standard of Perfection (SOP).

Each rabbit breed has a detailed standard that is available from the American Rabbit Breeders Association (ARBA).[1] The standards describe how much importance is given to conformation, color, fur quality, markings, and structure and body type. Far from being superfluous, each standard details the traits and qualities that make one breed unique from the another.

An SOP also has a lot of other information about rabbits, rabbit shows and health that can be helpful for the new breeder.

If you are just breeding crossbred rabbits for food production, an SOP may not be as important to you. However, setting standards and goals to evaluate and select your rabbits will allow you to improve your herd deliberately, instead of by making decisions randomly.

General Appearance

When evaluating a rabbit, Eric likes to stand back and evaluate the animal as it's just hanging out in its pen. We want a rabbit that is alert and aware of its surroundings, but that doesn't flip out and panic whenever it sees a human. Rabbits that are excessively nervous will not make good breeding stock. Either they will be poor mothers, or they will be burning so many calories worrying about everything that they will not be efficient meat animals.

A rabbit should be active, but not nuts. Hopping around the pen, grabbing a bite of food, a drink or chatting with the neighbors are all

fine, but be wary of a rabbit that panics and starts stamping any time you go into the barn. (Unless you are wearing the Coat of Death…see Chapter 6.)

When approaching a rabbit pen to evaluate an animal, look first at the overall general appearance. Does it have a shiny coat? Bright eyes? Alert expression?

A *shiny coat* is a good overall indicator of the general health of the animal. A dull-coated animal is one that is experiencing some sort of stress, either from a diet that is inadequate in nutrition, some sort of systemic illness, or a parasite load. The skin is the largest organ of the body for the rabbit, just as it is for us, and a shiny coat indicates the animal is in good health overall.

Look at the eyes. Are they bright, clear and free of discharge? Like with the coat, an animal with dull eyes and a dull expression is not feel-

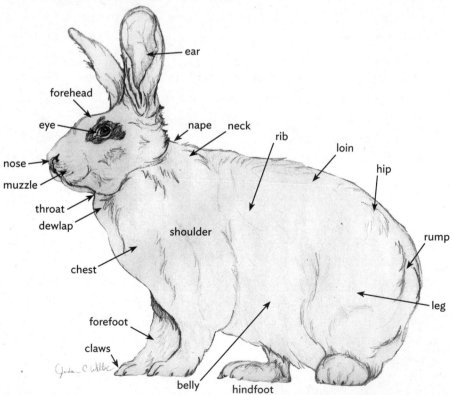

FIGURE 9.1. Rabbit features.

> ### A Word about Improvement
>
> Even if you are just breeding rabbits for meat for your own table, adherence to selection critieria will allow you to retain the best-growing, most feed efficient animals in your herd. This will maximize your time efficiency, cost-effectiveness, that number and quality of your herd. So don't discard taking a careful look at your stock, and selecting the best of the best to breed. Things like fur color may make little difference to the meat production, but make sure you have some standards to stick with as you work.

ing well. Also look to make sure the rabbit's eyes do not have discharge, which can be an indicator of snuffles.

If you have a table or workbench in your barn, take the rabbit to it, so that you can look it over easily.

Ears

Look at the ears. A rabbit's ears are of great importance, not just because they would look funny without them, but as an important part of heat regulation in warm weather.

FIGURE 9.2. This young doe has the alert expression of an active, healthy rabbit.

First, check for ear mites. Their presence isn't a dealbreaker, but a rabbit with ear mites will need to be treated. Anything that compromises blood circulation in the ear can pose problems down the road.

Ears should set well on the rabbit's head, neither too close together or far apart (unless the breed standard calls for either). Ears that are far apart can tend to droop as blood vessels expand in hot weather, and if they droop too much, air may not be able to circulate over them. Ears may droop a bit in the summer due to the increased blood flow, so be sure to note the temperature when you are evaluating them and account accordingly.

Ears should also not be too short, again for the reasons described above. One of the advantages to the three main heritage breeds we have worked the most with is they all call for large, sturdy well-set ears, which helped the animals tolerate Kansas's warm, humid summers.

FIGURE 9.3. American White Rabbit with good ear set.

Dewlap

Visually check the *dewlap*. This is the fold of skin under the rabbit's chin. Does will pull fur from this area to help make their nest. In older does that have had multiple litters, a large dewlap is not a problem, but in a young doe that has never had a litter, a large dewlap almost always indicates that she's gotten a little too fat, and may have difficulty getting bred. This would not be a sole reason to cull a doe in our barn, but if choosing between two candidates, the one with the more pronounced dewlap is going to get scrutinized a little harder.

Body Shape and Proportion

Run your hands over the rabbit. A good meat rabbit should feel, well, meaty—meaning there should be a good covering of flesh over the skeleton. You should be able to feel the backbone if you press more firmly, but it shouldn't be painfully obvious when running your hands over the rabbit.

FIGURE 9.4. This doe has a dewlap a little more pronounced that we like to see in young does that have not had a litter. She's a nice enough rabbit, so she will still get a shot at the breeding herd, but she will be watched closely for performance.

The rabbit's shoulders should be wide; our rule of thumb is that you should be able to fit at least one, hopefully two, fingers between the shoulder blades of a mature rabbit. Good wide shoulders are indicative of a rabbit that has plenty of chest capacity for organs to work efficiently; wide shoulders are also meatier.

The loin, which runs from the shoulders back to the hips, should be wide. The loin is the premium meat cut on the rabbit. A good loin should feel broad under your hands, and in no place feel pinched or drawn in.

There will be some differences between breeds in the width of the loin. The American Chinchilla (which has a more compact, commercial-type structure) will have a broader loin than the American Blue and White, but the American's loin will be longer, due to their mandolin-shaped body type. The American's loin starts a little further back from the shoulders than the Chinchilla's and is longer because of the longer body, but both should feel firm and meaty.

Run your hand over the body, down over the hips. Your hand should slope over the hips, rather than drop off. You will likely feel the hipbones, but they should have a good covering of flesh.

FIGURE 9.5. Rabbit with good width in front.

FIGURE 9.6. This rabbit is a little narrower in front than we like to see.

Rear and Tail

Make sure the tail is straight and not kinked to one side, which may indicate that they are not sitting evenly or have some structural issue.

Looking at the rabbit from behind, its hocks should sit well underneath neither turned in or turned out. Hocks turned in or out can lead to uneven pressure on the hind foot pads, and to sore hocks.

Looking at the rabbit from the top, it should resemble a relatively straight cylinder, not too narrow in the front, not pinched in behind the ribs, and in general shaped rather like a tube.

Weight

When evaluating rabbits as potential breeding stock, a good weight record is very handy.

We look for rabbits that maintain a steady rate of growth, and are constantly selecting those that outperform their siblings (when factors such as environment and feed access are consistent). Weighing growing animals regularly adds a step to the routine, but provides invaluable information for making breeding stock selections.

FIGURE 9.7. Good American Chinchilla profile.

FIGURE 9.8. American White Mandolin shape.

Posing for Show

When posing rabbits for show, the showman will position them so that the feet are under the hips more directly, and the rabbit has a rounder shape than he would naturally. We don't do this when evaluating our rabbits, and we don't show them—so none of them are used to being posed, and they really just don't like it. We evaluate them as they rest naturally, especially the long-bodied American Blue and Whites.

The rules of thumb for the three main breeds we work with most regularly are:

- American Chinchillas should gain close to 1½ pounds (0.7 kg) a month after weaning,
- American Blue and White, and Silver Foxes should gain 1¼ pounds (0.5 kg) per month after weaning. Foxes and Americans are both less compact than the Chinchilla, have a longer frame, and take a little more time to fill in that frame.

When to Breed

We are asked many times what the proper age is to breed the rabbits for the first time. We tend to rely more on weight than age. A rabbit becomes sexually mature at about 75–80 percent of their mature weight, which for our breeds is about 8 pounds (3.6 kg). We have some that hit there at about three months, some at four. Good weight records can help you target those individuals that are ready to get into production.

FIGURE 9.9. Chinchilla butts are the cutest.

CHAPTER 10

Breeding and Reproduction

Rabbits have a well-earned reputation for…well…breeding like rabbits.

Except when you want them to.

Conventional wisdom states that all you need to do is just throw rabbits together, and you will be swimming in rabbits in a few weeks. But often there are a few more challenges in store before you can start counting your kits.

Estrus versus Estrous

These two words are annoyingly similar looking and similar sounding, but they refer to very different parts of the same reproductive cycle. *Estrous* refers to the entire cycle from the first day of physiological activity by the ovaries, through the sexual receptivity phase, and then to the period where all ovarian activity regresses, and back to the point where it starts again.

Estrus refers solely to the period, usually right before ovulation, where females are receptive to mating. This is commonly called *being in heat.*

Distinguishing these two words has tripped up many a freshman Animal Science student, and in order not to make my editor's eyes bleed, for the purposes of this book I'm going to refer to the entire cycle as estrous, and the period of receptivity to mating, as either that or the more familiar being in heat.

Rabbits' Cycles

Rabbits do their estrous cycle a bit differently than other mammals.

Most livestock species have a cycle that is composed of a longer period when they aren't in heat, and a shorter (usually one to two days, depending on the species) period of receptivity to mating. The estrous cycle generally lasts 21 days on average, although it can vary from 18–24. Ovulation occurs at the end of the heat period, and mating is timed to make sure semen reaches the egg at the optimum time for fertilization. If the timing is off, or no breeding occurs, the follicles on the ovaries that produce both the eggs and the hormones that support pregnancy regress and the process starts all over again.

Rabbits are pretty much the opposite. They can have a period of up to two weeks when they will accept mating by the buck. But they do not ovulate until *after* mating has occurred. Rabbits are what is known in the animal kingdom as *induced ovulators*. Without the stimulation of mating, no ovulation will occur. Rabbits will remain receptive until the follicles regress.

Rabbits also have a much shorter period of non-receptivity, usually three to four days on average.

This is good news for the farmer. With cattle or sheep, if you miss the brief period of heat and don't get your bull in or the artificial insemination done, you have to wait 21 days to try again. A rabbit will usually give you an opportunity the next day. Usually.

Signs of Heat in Rabbits

Most does will not show a lot of outward behavior, but there are a few signs to look for.

Like most other livestock, a receptive doe can sometimes go off feed. (This is another reason to limit feed, and know exactly how much food she had the day before.) She can occasionally be more vocal, and a little bit more grouchy and territorial. If you think a doe might be in heat, be careful reaching into her pen. She might be a little defensive.

The most reliable way to tell if she's in heat is to look at the condition of her vulva.

A receptive doe will have a vulva that is dark pink and moist looking, while a nonreceptive doe will have one that looks pale and dry by comparison.

And the only way to accurately tell the condition of her vulva, is to pick her up, turn her over and take a good look.

How to Heat-Check

Of course, some does are less appreciative of being handled than others. But if you are not confident of your handling skills, this can also make the doe a little more nervous. Be calm and firm when heat-checking your does.

Grasp the doe firmly by the scruff of the neck. Pick her up out of the pen, and cradle her for a moment until she's calm, and then turn her over, still supporting her and keeping her close to your body. This should leave you with a relatively free hand to move the hair away and expose the vulva.

If you are smaller, have difficulty holding the big rabbits, or you have a reluctant participant, it's also possible to support her on your thighs or lap.

Either way, watch out for and respect those powerful hind legs.

At this point, you're probably asking yourself why you should even bother with this, won't the buck be able to figure it out?

Yes, that's his job. But if you continually put nonreceptive does in with him or does that aren't interested, or at worst, aggressive towards him, he's going to get discouraged pretty quickly and will be unlikely to want to put forth his best efforts in the future. This is especially true for a young buck that is learning the ropes the first few times. Help him out by offering him a doe who is also receptive.

FIGURE 10.1. Vulva of doe in heat.

FIGURE 10.2. Larger does can be a challenge to heat check. Holding the doe on your thighs can support her and give you the chance to see if she's in heat.

Training the Buck

We've heard it often enough: It must be easy to raise rabbits, right? All you have to do is throw two of them together, right? If you've read this far, you know there's a bit more to it, although rabbits have earned their reputation for prolificacy.

One of the most important steps you can take when starting out with a new group of rabbits is take the time to train your buck. Yes, mother nature has provided him with a good bit of hardwiring to get him off to a good start. But when starting a young buck, there are a few tricks you can use to make his first experiences pleasant.

- Make sure he's old enough. Different breeds mature at different rates, but the rule of thumb here is that he has to be at least six months old. His body weight also has some effect, they are generally sexually mature at about 75 percent of their mature weight.
- Eric has noticed over the years of tracking weights that the bucks will gain an additional ¾ lb (0.3 kg) after they start being used in the breeding herd. Some of this is just maturity, but increased testosterone also plays a part.
- Don't keep bucks and does side by side. Bucks will spray everything around them with their urine, and when you bring your buck a new doe, you want her to smell interesting.
- Always take the doe to the buck's pen. This keeps him from having to check everything out instead of paying attention to the doe, and his job.
- If you are sure your doe is in heat, but the buck isn't interested, take her out for 5–10 minutes to give him a reset. Letting them stay together can frustrate her, and she may try to ride him, which isn't how you want it to go; this can teach both rabbits bad habits.
- Bucks have an attention span of about five minutes. If he doesn't get the job done, or she isn't in heat, he can get frustrated.
- If you have older does, use your first-time bucks on them, and vice versa your experienced bucks and young does.
- If you have a pair of rookies, it's fine to reach into the pen and help get one or both of them into the right position. You want to be able

to handle and assist when necessary, but don't mess with them too much that they start to think they can't do anything without your assistance.

▸ We've said it before and we'll say it again: don't overfeed and let the bucks get too fat, especially if you have a small herd and they will go some time between matings.

▸ And above all, don't just leave them together and walk off. If one or the other gets frustrated, they can take it out on the other animal, and a cranky doe can really hurt an inexperienced buck. And, he can't read a self-help book to get his self-esteem back on track.

Time and patience are your best friends.

Seasonality and Lighting

Domesticated rabbits still experience seasonal variation in breeding. They retain some species hardwiring to have the majority of their offspring in sunny, warm months when available food is at its peak. And while some individual rabbits will show very little seasonal preference, if you are counting on year-round production in your rabbitry, you will need to set up a lighting program in your barn. This can be as simple as putting lights on a timer set to 16 hours of day length. (A timer will reduce the likelihood of forgetting to turn the lights on or off.) Inconsistent lighting will do more to thwart your efforts than no light at all.

Your rabbits may also experience a little slump in fertility and reproductive rate in the summer months, if you live in a climate where temperatures regularly gets above 90°F (32°C).

Doe Body Condition

It's important to not overfeed your breeding stock, especially young does (and bucks) before their first breeding. It takes very little extra feed to get a doe overconditioned, and then it becomes very difficult for her to conceive.

Rabbits do not put on intramuscular fat, nor will they develop a nice "fat cap" under the skin over their carcass. Any extra calories stored as

FIGURE 10.3. Doe with visceral fat. This comes from overfeeding, or when animals grow too fast.

fat will go straight to the body cavity, surrounding the internal organs. Once that fat is in a doe's body, it becomes difficult for ovulation to occur and for her to breed. We have processed over the years several does that would not get bred no matter how many times they were serviced by the buck, and these does were almost always full of visceral fat.

And at that point, there isn't a diet in the world except starvation that will remove visceral fat. It's much easier to just not overfeed them in the first place.

It helps to think of breeding stock as athletes. No one can get great athletic performance by pigging out at the all-you-can-eat buffet on a regular basis. It just doesn't work that way for humans, and it doesn't work for rabbits.

Encouraging Successful Breeding

There are very few absolutes when dealing with animals, but here's one: ALWAYS take the doe to the buck's pen for breeding.

Does can be very territorial, and bringing the buck to her cage is an invitation for her to be aggressive towards him, possibly injuring him or at least ruining the mood.

A good, mature, experienced buck will be all business. He may take a few nudges at her to test the waters, but he will be focused on getting the job done. Mating should occur within a few minutes, based on the experience of your buck and doe. It might take a younger pair a bit longer to figure things out, but they should at least be interested in one another.

Stay there until mating does or doesn't occur. Sometimes if he's slow to go to work, it's tempting to leave them together for an extended period of time, but this is inadvisable. Mating might occur while you're gone and you will not know the outcome, or she might get tired of his advances and fight with him.

When rabbits fight, they use their hind legs and claws to attempt to eviscerate one another by clawing and scratching at the underbelly, which can also put a doe's flailing claws right next to a buck's testicles. Rabbits can effectively castrate one another, which is a dismal end to a promising breeding program.

Above all, do not leave the doe in the buck's cage unsupervised in the hopes that they will "work it out." At best, no mating will occur—but you won't know it until 30 days later when she doesn't have a litter. At worst, the doe and buck may fight. An aggressive older doe can intimidate a buck to the point where he is reluctant to even attempt to breed.

Now, a little chasing is a normal part of rabbit courtship, so don't be alarmed if they engage in that kind of foreplay. But watch for actual fighting, and don't hesitate to separate them and try again later. Sometimes a few hours can make all the difference.

If you have a doe that you are positive is in heat, but she won't hold still for the buck to mount her, you can reach into the cage to hold her still. Grip the skin on her shoulders and hold her for a few seconds until the buck mounts her. Moving around too much is more common with young, inexperienced does, and usually they get the picture pretty quick.

When successful mating occurs, the doe will raise her hindquarters for the buck. He

FIGURE 10.4. Eric holds the doe for the buck so he can get the job done.

If you feel like your presence is distracting to either the buck or the doe, it's OK to step away for a few minutes, but don't step so far away that you can't see them at all. Step over to the corner, finish your coffee, and pretend to check your email on your phone (whatever it takes to assure them you aren't going to bother them), but keep an eye on them.

FIGURE 10.5. Buck rabbit falls away from doe as mating is completed. This is hard to get a picture of; it can happen very quickly!

will breed very quickly with very rapid thrusts, and when ejaculation occurs, he will lose his footing from the force of the thrusts and fall over, sometimes with a little cry.

If the buck doesn't fall over, no mating has occurred.

Care of the Doe after Breeding

We always weigh the doe after breeding, and treat her ears with a couple of drops of mineral oil. This helps keep ear mites at bay, and the weight gets recorded and becomes a part of her permanent record. We also weigh again at weaning, and expect a good production doe to, at the very least, maintain her weight through lactation.

After weighing her and treating her ears, return the doe to her own pen. Make a note on the cage card of the date, and take a minute to calculate the date that she can be palpated (see "Confirming Pregnancy" later in this chapter), and the expected due date. And it's a good idea to make a note somewhere else, whether it's on your phone's calendar app, a text to your spouse, or a paper calendar.

The doe does not require any special care at this point, and resist the temptation to start "feeding her for eight." If she is in good condition and at a good weight, a pregnant doe will not need any extra calories until she begins nursing the kits. In fact, extra feed at this point can lead to kits growing too large to be born easily, and can cause the mother to overproduce milk, leading to *mastitis* (inflamed udders). Wait to make any dietary adjustments until after kindling, and when she is in full lactation.

Care of the Buck

The buck is the most important member of your herd. Since he will be mating with multiple females, his genes will be the biggest influence on your production and the quality of offspring.

A good buck should be bright-eyed, and eager to work when you bring him a doe.

Breeders with small herds can run into the problem that, if the buck is not used frequently and especially if he is overfed, he can become fat and lazy quite easily. Bucks who are fed too much frequently have a much lower libido and may not even be interested when presented with a female in quite obvious heat. Bucks should never be allowed to get fat. A single 4-ounce (113 g) feeding per day of quality pellets will be enough to meet his needs. If you choose to feed foods other than pellets, food quantity might be a bit trickier to balance.

A single buck has the ability to mate with many females before he suffers a loss of potency. At one time the rule of thumb was that a single buck could serve 10 does, but some research has shown that they can service even more.[1]

Decide what is going to work best for you. If you have access to replacement stock, fewer males might be the way for you to go.

In any case, it is a balancing act between keeping your buck fit and in breeding condition, and overwhelming yourself by breeding too many litters.

Bucks can also experience seasonal sterility in the summer. Temperatures over 85°F (29°C), especially

A rabbitry that is breeding for *conservation of rare breeds* will need to keep more bucks to make sure individual lines can be kept and monitored to keep the overall herd viable.

A herd that is strictly for *meat production* and not keeping replacement stock can get by with a single buck, provided you have a resource to replace him when and if it becomes necessary.

when it remains hot at night, can cause rabbit bucks to temporarily become sterile. Sometimes, this sterility can become permanent. We run a cool cell in the barn our bucks are housed in during the hot summer months, to reduce the effects of the heat as much as possible.

Confirming Pregnancy

With other species of domestic animals, a 21-day heat check is often the first concrete sign of pregnancy. With rabbit, however, their gestation is two-thirds complete at that point!

In many rabbitries, the practice is to just wait until that kindling (rabbit birth) date comes and goes to confirm if the doe was pregnant. If your goal is rabbit production, that strategy is inefficient.

Another common practice is to try a test breeding at about two weeks after the initial mating. However, this strategy has its drawbacks also. Some does will breed even though they are pregnant, and unless you pay attention to the original due date as well, she could kindle then and, if you aren't prepared with a nest box, she could lose the litter.

FIGURE 10.6. This photo shows where Eric places his hand when palpating. It looks like he is really squeezing the doe, but he actually just lets her belly fall into his hand.

FIGURE 10.7. A doe that is cranky can be palpated in her pen. The positioning of the hands is the same, but she has more support and is confined a bit better.

Palpation

A better, more accurate solution is to palpate does to confirm they're pregnant. This is a relatively simple procedure and won't hurt the doe, but it does take a bit of practice to be accurate.

Restrain the doe with by the scruff, and slide your other hand under her abdomen. Let the weight of her belly fall into your hand. Slide your fingers and thumb along either side of her abdomen, and you should feel little round objects (about the size of grapes) if she is pregnant.

At day 11 or 12, a person experienced at palpation should be able to feel the uterus, and the small, marble-sized kits. Until you've gained a lot of experience palpating, day 14 is going to be a much more realistic date to check. Eric can palpate at day 11 and tell with a high degree of accuracy how many kits the doe is going to have. Me, I settle for pregnant or *open* (not pregnant).

You will have to palpate a lot of does to get the ability to count kits with any degree of certainty. It's possible to mistake kidneys for kits. This is where it's helpful to practice regularly. Kits will change and grow; kidneys do not.

Practice on a couple of does you know are open, and then as you practice with bred ones, you can at least begin to tell the difference between an empty uterus and a full one.

Most people are concerned about hurting the doe or her litter when palpating, but as long as you are not intentionally trying to squeeze her innards into mush, a healthy doe with a healthy pregnancy will not suffer ill effects.

Palpating might seem like an unnecessary task. However, non-productive doe days can tank the profitability of the rabbitry. Even if your goal is not to sell the fryers for a profit, having fewer animals can still reduce income. Even if your goal is to just have a freezer full of meat by winter, a two-week loss in production can set things back further than necessary.

Can you guess wrong? Of course. It still pays to keep a close eye on her at the first due date.

Watch for signs of impending kindling (heavy belly, nesting behavior). If you palpate close to the due date, you should feel a heavy, full belly. Go ahead and put the nest box in: better safe than sorry.

CHAPTER 11

Kindling

Nest Boxes and Maternal Behavior

Does may become restless as their kindling date approaches. At times they will go off feed, and a doe that normally cleans up her feed and suddenly starts leaving some might bear closer watching.

The nest box and bedding should go in with the doe at day 28. Rabbits have a 31–32 day gestation, so providing the nest box a bit early gives her time to make her nest and prepare without leaving it in long enough for her to decide that the nest box makes an even better toilet. Seeing the box and the bedding will often trigger her into a home remodel, and she will begin getting things arranged to her liking.

You may also see her gathering her nesting material, and beginning to pull fur from her chest and belly region. (There is a hormonal trigger during the early stages of kindling that causes those hairs to be released more readily, so don't be alarmed that she's causing herself a lot of pain.)[1]

Normally the doe will take that fur, mix it with the bedding you have provided in the nest box, and make a nice, comfortable nest, called a *form*.

Sometimes a doe will get carried away with the fur and actually have too much, especially in the summer months. If she has overdone it with the fur, you can pull some of it out and store it in a freezer bag, in case you have a emergency when another doe fails to make a good nest sometime in the future.

At some point the doe should disappear into her box, and that is the time to leave her alone. In general, does will kindle in the early morning. In an ideal situation the healthy, wriggling litter will already be there when you check on her in the morning.

FIGURE 11.1. Doe, with a mouthful of her own fur, gets ready to make a nest.

FIGURE 11.2. Good winter nest, with deeper bedding.

Rabbits do not intensively mother their young; does will generally seem to be completely ignoring the kits. Remember that as a prey species, a rabbit doe's instincts are to do nothing that will draw attention to the nest.

Some does make nests that can truly be considered works of art: warm, round little depressions in the nest box lined carefully with fur and more than capable of keeping healthy kits warm in all but the most bitter of weather. Inside the nest box temperatures can run up to 15°F (27°C) warmer than outside.

Kindling and After

Kindling generally happens quite quickly: one minute you have an empty nest box, and the next there is a warm wiggling litter of new kits. It takes about 15 minutes start to finish for the doe to have her babies. They do not clean them up like other mammals do, but if given the opportunity, they will eat the *placentomes* (the little parts of the placenta that come out after the kits). Rabbits will also not pass a single placenta like cows or sheep, so don't be alarmed if there is no evidence of the birth other than the litter.

Kits also almost never require assistance to be born. It would be physically impossible to try to go in and assist her anyway, and you would likely do much more harm than good.

After the doe has kindled, check the nest. Does will not reject their offspring if you touch them, but be sure you haven't petted the dog or cat beforehand. If it still concerns you, take a handful of the bedding and rub your hands with it, or rub her for a minute or two before checking the nest.

Healthy kits will be warm and wiggling in the center of a fur-lined nest. Kits do not seem to compete for the warmest spot in the nest, but rather seem to rotate and take turns working from the outside to the inside and back again. Count how many she's had at this point, and make a note on her card.

FIGURE 11.3. Good summer nest, with shallower bedding.

We check nest boxes after kindling, to get a count, and then every day after that until the nest box is removed. This allows us to make sure all the kits are thriving. If one or two die, the doe will push them out of the nest and into a corner of the box, where their bodies can decompose quickly. Take the time to remove any dead kits as soon as you find them. If the nest box becomes particularly foul, it can cause the doe to not want to enter it, and her kits could starve. A nasty nest with decomposing kits is also a haven for flies and maggots, and can lead to bacterial challenges that kits cannot overcome.

Does do not spend a lot of time in the nest, only entering a couple of times a day. (Some will also jump in to check things out after Eric has checked the boxes, but they exit right after they've approved of his job.)

Rabbits will only nurse their kits once or twice a day, and only for a few minutes. Their milk is of necessity high in protein. A short time before she is due to nurse them, her kits will begin wriggling around in the nest, warming themselves up and getting ready to make the most of their one shot at a meal. At this time, if you check the nest, you will notice how active they are, and if you reach in to the nest, they may pop around and squeak.

The doe will jump in the nest, lower herself over the kits, and they will feed. Kits do not seem to have a teat preference, and will latch on to any teat that they can. If some kits do not get a chance to nurse, there

FIGURE 11.4. At one week, kits are covered with baby fur and show small, distinct ears.

is no second opportunity. The doe will not go back until the next time, and does not check to see if everybody got all they needed. Kits can survive until the next nursing opportunity, but if they get outcompeted by their bigger siblings again, they are at risk for starving quickly after that.

This nursing behavior can take a bit of getting used to if you are more familiar with species that nurse multiple times over the course of the day, or even several times an hour. But if the kits are warm and have full tummies, the doe is doing her job.

Most does will only exhibit mild concern when you check their nest. This is another advantage of a tote-box nest box. It allows you to pull it to the cage opening and reach in without her in the way. Some does, however, can be more defensive about their nest, so use caution until you are more familiar with each particular doe's behavior. While a little protectiveness is welcome, outright aggression is not. Give her the benefit of the doubt if she's a first-time mom right after kindling, but if she continues being overtly hostile, you might want to reconsider her future in your breeding program.

Kindling Problems

Litters of rabbits seem either to do really really well or really really poorly. Unlike other livestock species (where you have a larger window to intervene, assist in a birth, bottle-feed an animal and, have a larger role in the offspring's survival), the relative speed with which a litter of rabbits can go downhill gives you little time or ability to act.

Most commonly, one day your litter will be fine, and the next...not. Generally, the first one to three days are the most critical time for a new litter of rabbits. Below are a few things to watch out for with every litter.

No Milk

Unfortunately, often the first sign of no milk is a litter that begins to starve. If you have other does that have similar litters you can foster kits from the poor litter to the stronger one, but be aware that if the starving kits are not strong enough to compete with the other litter, they will likely die anyway.

Lactation is a fairly complicated physical and hormonal process, and stress can be enough to short-circuit things. Make sure she's in a calm environment, without a lot of traffic or noise to startle her.

Lack of milk production is also more common in first-time mothers, whose mammary systems may not be fully developed yet.

Nutrition can also play a part. Lactation is one of the most nutritionally stressful events in a rabbit's life. Make sure the doe is on an adequate diet (around 17 percent protein),[2] and that she has plenty of fresh, clean water to drink.

Mastitis

On the other hand, overfeeding a doe can lead to another problem: *mastitis* (caked breast). Too much of a too-high-protein feed can cause a doe's mammary system to go into overdrive. Rabbits only nurse their kits once or twice a day at most. If the doe produces more milk than the kits can consume, her udder can become engorged and inflamed, and that can cause her milk production to shut down. (For tips and methods of feeding lactating does see Chapter 12.)

If the udder feels warm to the touch, you can place a cool damp cloth on the udder to help reduce inflammation and stimulate letting down of milk, and then place the doe back with the kits to nurse. The challenges here are that the doe might not appreciate your cooling cloth efforts, and you may not know when exactly she will nurse her kits.

Fostering the kits can be an option, and they can be replaced on the problem doe with kits that are a little stronger and older (a little…she's not stupid), but your best bet is prevent mastitis by not overfeeding in the first place.

Cannibalism

It's truly horrifying to come in and find parts and pieces of a litter scattered all over the pen and the nest box. In our minds we can't conceive of what would cause a rabbit mother to do this, but again, keep in mind that she will sacrifice her litter if she doesn't feel she can raise it successfully. And, in her mind, eating the evidence to elude predators is a good strategy.

Most of the time, cannibalism will happen with first-time moms, and then most of them go on to raise litters successfully. Again, step back and evaluate the environment. Loud noises? Dogs barking? Kids running in and out? Some does will calmly carry on in the face of all sorts of distractions, but for many it is too stressful and they can't manage.

In our barn, we have a three strike rule. We know our environment is consistent. A first litter, we can put down to just being a nervous new mom. A second litter, she's on the short list. Unless she has some very desirable trait we are trying to capture, she may be culled at that point. Third time? She's a poor mom, and headed for freezer camp.

Dystocia

Dystocia is a word that encompasses all sorts of problems with actual birthing itself. It can include malpresentation of the kit in the birth canal (which is not common) or too large a kit due to small litter or overfeeding. Fortunately, dystocia is not common in rabbits. Most of the time they can pass their babies quite successfully, even larger ones. We have seen some single kits as big as two normal-sized ones pass successfully.

It is also very difficult and inadvisable to try to reach into the doe to try to straighten something out. Give her some room to move around, and see if she can't sort things out herself.

Stillborn

Finding stillborns is common. It is sad to lose an entire litter; if that happens, take a good look at the environment and see if any adjustments can be made. Stillborns are more common with does that go too far past (2–3 days) their due date. Sometimes does will go a day or two longer in hot weather and everything is fine, but much past that and problems may occur.

Rabbits rarely have spontaneous abortions. If she does, then there is some underlying medical issue that has more than likely caused the problem. You can rebreed her, but it may be necessary to cull her.

Occasionally, you will find what looks like a mummified fetus in the nest box. These are not common, and generally the result of an embryo that has died at some point during gestation.

Early Nest Box Death

This is often related to a problem with the doe, but can just as often happen because a nest box has become foul. Kits have no real immune system in the first few days and are susceptible to any challenges. If the doe has urinated or defecated in the nest box, the environmental challenge will be too much for the kits.

Keep the box dry, and the bedding clean and fresh. Examine the corners of the box for any deceased kits tucked away there. The doe is doing her best to keep things clean by moving that dead kit away, but once it starts to decompose, it can foul the whole box, causing her to abandon the litter.

Chilled Kits

A chilled litter will feel cool to the touch, contrasted with a healthy litter that will feel quite warm by comparison. Chilled kits are also extremely lethargic, sometimes so much they can be easily mistaken for dead. If you check the box daily, as you should, and notice kits that are alive but chilled, this is a sign that there are problems with the litter. Chilled kits cannot feed, and conversely kits can become chilled because they are not being fed.

Extremely chilled kits need to be warmed up as soon as possible, and fostered onto another doe if at all possible. There's about a 50/50 chance that this litter will make it, but it's 100 percent sure they will die without intervention.

Live Young Scattered on the Wire

First-time moms can seem quite perplexed and confused by these wriggling things that are coming out of their bodies, and not seem to have a clue what to do about them.

If you find kits outside the nest and on the cage wire—and they are warm to the touch, gather them up and place them in the nest, hollowing out a little area to contain them if the doe has not done so herself. Then leave the doe alone to figure things out. She either will or she won't, and fussing over her will just convince her that those little squeaky things are bad news, and she'll abandon the entire litter.

If the kits you find on the wire are chilled, warm them up before putting them back. Even if she were to go right in to feed them, chilled kits cannot nurse or digest adequately and they will die regardless. Chilled kits can be warmed up by putting them in a nest with another healthy litter (make sure you can tell who's who) and then returned to the doe, or by putting them under a *kit warmer* (hanging an incandescent light bulb over a small bucket with some bedding in it).

Your goal is to warm the kits up to body temperature, not just warm them to the touch on the outside.

Once you've looked after these kits, take time to evaluate the whole environment of the pen. Can anything else be improved?

Feeding the Lactating Doe

We maintain the 4-ounce (113 g) feed ration for a doe through the first 10 days after kindling. Since she's now lactating, the temptation is strong to up her feed immediately, but an increase in diet before the kits are large enough to consume extra milk can lead to mastitis and the doe ceasing to produce milk. Or the kits can develop *scours* (diarrhea) and potentially die off rapidly.

> Do not use a heat lamp bulb unless you are prepared to continually monitor the temperature of the kits. A few minutes are all it takes to raise that temperature in the bucket to a fatal level for the kits.

We continue with that 4-ounce scoop until day 10, at which time we offer a second scoop. The kits are now much more active and consuming a lot more milk; they are less likely to have digestive problems, and they are able to keep the doe nursed down enough to prevent mastitis.

Originally we raised the feed level at day seven after kindling, but still had issues with scouring. This has nearly disappeared since we now wait until day 10 to up the doe's feed.

Kit Growth and Nest Box Management

Rabbit kits are born completely helpless, with their eyes closed. At about day 10, their eyes will open and they will begin exploring the nest box.

As they grow, they will be urinating and defecating in the box more frequently, and the bedding will need changed more regularly.

Around days 14–17, they may start coming out of the box to check things out. Occasionally one will get dragged out when the doe exits after nursing, so make sure the smaller ones can get back in where they belong.

Eric will often remove the nest box at day 17. At this point, unless it is bitterly cold, the box does not provide any extra warmth, and it will

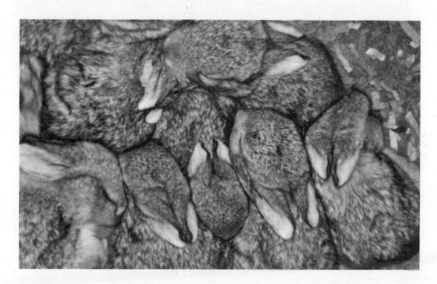

FIGURE 11.5. At two weeks, these kits have coarser fur, distinct ears and eyes.

be beginning to stay damp from urine. You can take the nest material out of the box and place it in the corner of the pen for any smaller ones until they become adjusted to the wire floor. They will work the nest material down through the floor as they move around. It's not necessary to put anything else in for the kits. They will be healthier and drier without constant exposure to wet bedding. They will huddle together with the doe and each other and manage just fine.

If you must add something until the young can walk on the wire, a handful of grass hay or bedding will be best.

Losing litters is most often related to management. If you are continually having problems with litter loss, take a good look at your management practices. Assuming you purchased stock from a reputable breeder who focuses on production qualities, management is the most likely culprit, but it is also the easiest to change.

Fostering Kits to Other Litters

Because we have litters born continuously and raise many visually different breeds, we are able to foster kits from one litter to another with relative ease.

Fostering kits can help save disadvantaged ones from starving out and dying. Does will readily accept kits from other litters, and don't seem to really count or care who they are feeding. The kits need to be of as similar a size as possible to prevent disadvantaged ones from getting starved out again.

When fostering, make sure the kits being fostered are not sick, wet, or suffering from scours. Not only will they likely not survive, but they could possibly infect the good litter where you put them. It is also possible that the smell of a sick kit could cause a doe to reject both litters.

Moving larger kits out to a new litter can also be helpful. Larger kits will stand a better chance of not getting outcompeted by their new nest mates, and the smaller ones that are left behind will hopefully benefit from having less competition in the original nest.

Brush all the old bedding off the kits that you are going to move, and get a large pinch of bedding from the new nest. Rub the bedding from the new nest on the kits being fostered, and place them in the new nest. It will help the transition if all the young smell the same.

We've heard about people putting something else with a strong smell on a doe's nose to confuse her when fostering, but we have never found this necessary, and it would most likely just annoy the doe.

If you are breeding with the intention of keeping stock, it is helpful to have does from a couple of different breeds as potential foster mothers; otherwise it can be nearly impossible for you to be able to tell whose kits belong to which doe. If the litters are of the same breed, you can track offspring by giving them a tattoo, or an ear punch, which is really only feasible for larger kits. A permanent marker can be used to make a mark in the ear of the kits you want to track, but be warned this will need to be reapplied almost daily, as rabbits work really hard to keep their ears clean.

If you are crossbreeding and have does and litters of different colors, this can help with the identification; but if you are just breeding for meat and have no interest in tracking parentage, it doesn't matter who is who.

Litter Loss

Unfortunately, it happens. If your doe does lose her litter at birth or soon after, wait 72 hours for everything to clean itself out and get back to normal, and then rebreed her.

First-time rabbit mothers often seem to be clueless enough that you wonder how the species ever evolved in the first place. Do not be discouraged. It helps to realize that in that rabbit's mental hardwiring, her litter is disposable.

As mammals who invest a long time in gestation, and an even longer time raising our infants to a stage where they can fend for themselves, we humans, as a general rule, never consider abandoning our offspring to start over when conditions improve.

But as horrifying as this notion is to us, for the rabbit it is just part of the genetic makeup she inherited from her ancestors.

▸ Rabbits are able to have multiple offspring.

▸ Rabbits can have multiple litters in a year.

Both of these traits enable them to move on from an unsuccessful litter and be ready to try again immediately.

What motivates reproduction in any species? It is the desire, although not a conscious one, of the genes to reproduce and pass themselves down to the next generation.

As a prey animal, if the rabbit doesn't live long enough to reproduce, then her genes don't have that opportunity. So, at least speaking from the standpoint of her genes, it's in that rabbit's best interests to start over when the opportunity is more favorable.

Contrast this with another mammal, the elephant. The elephant has a 22-month gestation, gives birth to a single baby, and will spend several years nurturing that offspring before she will even be ready to reproduce again. Over the course of her reproductive life, she may only get at best five or six chances to pass on her genes. She, of necessity, has to be much more invested in the success and survival of that single baby, at the expense of her own life if necessary.

So while it's the last thing we humans would consider with our offspring, be forgiving of the rabbit if that first litter is not a success. Step back, honestly evaluate if you have sincerely done everything in your power to create a good environment for her, and try again. Nine times out of ten, a few simple adjustments in the environment, and a little experience, will put most does on the right track.

CHAPTER 12

Milestones and Management— Birth to Harvest

It is true that each litter is different, and will provide different lessons along your learning curve, but every litter will encounter several consistent milestones, and there are also several routine management steps that will enhance your chances of success along the way.

Feeding the Litter

When going into kindling and lactation, a doe should be in good enough body condition that her internal reserves will allow her to feed her kits without pulling her own condition down. After 10 days, the kits' growth becomes much more rapid, and their demands on the doe increase. This is the time to increase her feed; now that her lactation system is up and running, the kits can keep up with the amount she produces. Once the kits are 17–21 days old, we up the feed again, less for the doe as much as encouraging the kits to eat as much as they want. By that time the kits will be eating well and taking the bulk of the feed for themselves.

How Rabbits Grow and Develop

Rabbits are born *altricial*, meaning they are born with their eyes closed, and are completely dependent on their mother's care.

They grow rapidly however, and in about 10 days, their eyes open and they begin exploring the world of their nest box. It is critical to

TABLE 12.1. Litter timeline and milestones.

Day 1	**Litter is born** ✓ Count kits and remove dead ✓ Weigh doe and record ✓ Put mineral oil drops in doe's ears to prevent mites ✓ Clean out any bloody or wet nest material
Days 2 and 3	**This is the most critical time for your litter** ✓ Check for chilled kits daily
Days 4 through 9	**Check litter daily** ✓ Remove any dead ✓ Check for weak or chilled kits ✓ Remove damp nest box material
Day 10	**Increase doe's feed** ✓ Kits will start to open their eyes
Days 11 through 16	**Continue checking nest** ✓ Kits will begin to explore and may get out of the nest box
Days 17 through 21	**The nest box can be removed** ✓ Day 17 in warmer weather ✓ Day 21 in colder weather ✓ Adjust water bottle to kit height
Day 21	**Kits will be eating solid food**
Days 28 through 35	**Kits should be eating well** **Weaning can be done at this time** ✓ Weigh doe at weaning ✓ Sex kits and weigh litter ✓ Put mineral oil drops in doe's ears
Day 35	**Doe can be rebred**

TABLE 12.2. Feeding program for the doe and the litter.

Up to kindling	4 oz (113 g) per day
Kindling to 10 days	4 oz (113 g) per day
10 through 21 days	8 oz (226 g) per day
21 days through weaning	Full feed (what they will clean up from one feeding to the next)
Weaned kits	Full feed

help keep the nest box clean and freshly bedded at this stage; the larger and more active the kits are, the more they will urinate and defecate. Change out a handful of bedding daily or every other day, making sure to get wet matted bedding out of the corners. Wet, dirty bedding just serves as a breeding ground for bacteria, and can also harbor maggots.

The doe will not reject her babies with your daily care and checking, but if it concerns you, take a handful of bedding and rub your hands on it before checking the box and the kits.

Remove dead kits promptly. Again, they are unsanitary and just a breeding ground for bacteria and pests.

After about day 17 you can remove the nest box entirely. Especially in the summer, the nest box just becomes a litter box, which in turn becomes a nasty bug and cootie farm. Also, giving the doe more room to stretch out and cool off will be to her benefit. It also allows for better airflow around her and the kits. The kits will look tiny and vulnerable in the pen, but the fresh air will do them good.

Make sure the water source, whether it is a bottle, automatic drinker, or pan, is at a level the kits can reach and drink from easily. The doe will be able to drink from it even if it is lowered. Kits that are able to drink will get on solid food faster. Getting on solid food is crucial to them growing well and removing lactation pressure from the doe.

At four to six weeks, the kits will be ready to be weaned.

Weaning

Remove the doe from the litter. Leaving the litter in the same pen keeps them in familiar surroundings, and they will not be exposed to a new environment/set of challenges.

If you are using water bottles or automatic drinkers, make sure the drinker nipple is adjusted for the kits' comfort and ease. If kits don't get enough water, their feed intake will be slowed, and they will not perform as well. If using cups, make sure they are kept clean, and that fresh water is always in them. Kits (and adults) are notorious for using their water cups for at best, face washing, and at worst, a toilet. Keep fresh water readily available to newly weaned kits, even if you have to water more frequently, especially in the beginning.

Check the doe's weight, and condition. If her condition is good, and her weight has stayed the same, go ahead and rebreed her. Don't wait too long to rebreed; rabbits thrive reproductively when kept in a well-managed cycle of breeding and kindling. This is not to say she should be pushed past the point where she begins to lose condition, but keeping her from "going stale" will increase your chances of her breeding successfully. A good healthy doe can have four to five litters a year without difficulty.

Leaving your kits with their mother will not increase the kits' growth significantly. By the end of four weeks, her lactation curve has tapered off, and the kits, even if they attempt to nurse, will not be getting enough to justify aggravating the doe. And please don't leave the kits with their mother so long it becomes difficult to tell who is who. This is a sure way to get someone bred you don't really want her bred to.

If space allows, weaning is a good time to split up male and female kits. An adjacent pen can be used to split the litter, that way you are not moving one group too much or too far, and stressing the whole litter. Keep in mind that young kits will grow fast and become sexually mature relatively quickly. We advise splitting them up as early as possible to avoid accidents. And, as a side note, check them again in a couple of weeks. The "sex-change fairy" has visited our barn on more than one occasion, and it never hurts to be sure you've correctly sexed your rabbits.

Age at Harvest

Rabbits can be harvested right after weaning, although at an average 50–55 percent dress out, harvesting them that young doesn't offer much usable yield. There seems to be a sweet spot at about 12–14 weeks (for heritage breeds, some commercial rabbits and crosses will get there faster) or at a weight of about 5½ to 6 pounds (2.5–2.7 kg). This gross weight will generally yield a fryer with a carcass weight of about 2¾ to 3¼ pounds (1.25–1.5 kg). Letting rabbits get bigger isn't a problem, but the feed efficiency goes down the longer you feed them. If feed costs aren't a concern, or you want a larger carcass, go for it. The growth curve will slow down quite a bit from 16–18 weeks onward, so take that into consideration when planning your harvest. Letting rabbits get older won't hurt anything but your feed bill, but their most prime carcass age is before one year of age.

Older rabbits, such as unproductive does, can be harvested at any time. Their carcass size will be larger, but meat tenderness will go down after about a year to a year and a half.

Rebreeding the Doe

How do you know when it's time to rebreed?

Amazingly enough, rabbit does can be rebred within a couple of days after kindling. Many commercial rabbitries can produce eight litters a year from a single doe, which necessitates does being rebred within a week or two after kindling. Such a doe will barely get one litter weaned before the next is being born. She will require a high plane of nutrition to keep this up, and it will be difficult for her to maintain good body condition.

As you would expect, this intense of a production schedule can take a toll on the doe, and her reproductive life will be much shorter overall.

Managing for three to four litters per year is a much more reasonable schedule, both for the condition and well-being of the doe, and for keeping her reproductively fit.

Giving a doe too long a break between litters can make it more difficult to get her bred when you decide it's time. In some cases, their ovaries seem to shrivel when inactive.

Based on day length, there is also a seasonal drop in production. This is much more pronounced in the wild rabbit, and has been reduced greatly in domestic rabbits. But there may still be somewhat of a seasonal slump in production during the winter months. This can be solved by adding lights to the barn, and maintaining a light schedule similar to summer months (see Chapter 10).

Take a look at your doe and her litter. If she has a large litter (nine or more kits), give her a much needed chance to rebuild her energy stores by giving her that extra couple of weeks to recover.

If her condition or weight has started to slip, she needs a little more time. A good scale and a routine of weighing the doe at kindling and at weaning can give you an idea of how much the litter has drained her.

Do not attempt to just feed her more and extra to try to pack weight back on her quickly either. The odds are she will lay on unwanted fat as well. Continue to feed her a good ration consistently, and if everything else is all right, her condition will improve quickly.

> Inbreeding coefficients tell you the probability of inheriting two copies of a single allele (different gene) from an animal that is on both sides of the pedigree. If it is a good allele, problems will be minimal; if it is a bad allele, problems can arise. There are many factors that go into the genetic makeup of an animal, don't get too hung up on just one.

Record Keeping

Even if all you desire to have is a few rabbits to put meat on your table, take the time to keep good records. Human memory is fallible. A simple notebook, cage card, or electronic note on a computer or other device can save you a lot of head scratching and possible mistakes when evaluating rabbits.

There are several software programs out there that will track matings, show information, color genetics, and pedigree data. One of the most popular is Evans Rabbit Register, which is a good, easy-to-use program, and if tracking show data is a primary goal, this one is used by most of the rabbit folk we have come across. I can't speak for any of the other programs, and I'm sure many of them work quite well. But I can say I've always been pleased with Evans, and the customer support provided is top notch. The Evans program tracks pedigree information easily and has several search features which can help track down an elusive rabbit. The program also calculates inbreeding coefficient functions.[1]

Remember the old adage: "It's linebreeding if it works, inbreeding if it doesn't."

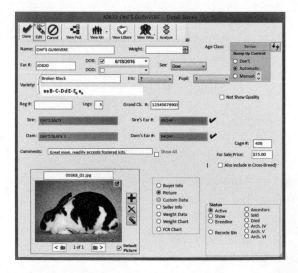

FIGURE 12.1. Digital record—single rabbit.

If you only have a few rabbits and don't plan to do much showing, you can always keep pedigrees by hand or in a generic electronic spreadsheet. However, you won't be able to calculate inbreeding

FIGURE 12.2. Digital record of pedigree line.

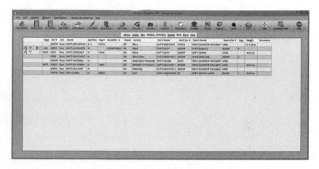

FIGURE 12.3. Summary table—rabbits in one pedigree line.

Eric's Little Black Book

In the early days, Eric kept records in a little pocket calendar book. With only a handful of does, it was pretty simple to keep track of who was who, and who was bred to whom. Eric kept everything in it clipped tightly with a little binder clip, and usually put it on his desk at the end of the day.

Well, one day in a fit of domesticity, I grabbed up his clothes to do laundry.

Yep, you see it coming. His little black calendar book was in a jeans pocket, and a quick pocket check didn't find it.

When I pulled the laundry out of the washing machine, I started noticing all these paper shreds in the machine. Lots of them. (Cue overly dramatic music.)

All that remained of the little book was what was held together with the binder clip. Half a year's worth of records.

Telling him that was about as bad as telling your dad you wrecked the car.

So, suffice it to say…always have a backup.

coefficients very easily. Whether you choose paper or electronic methods of record keeping, plan on some sort of backup system. Accidents do happen.

Each rabbit should also be identified by a cage card, which is just simply a card hanging on the animal's cage that can carry information about matings, due dates and any other pertinent information about that rabbit.

In the beginning, we had a problem with humidity and dust making paper cards unreadable in the long term. We solved that problem by getting packs of plastic index card dividers at the local office supply place, and we now use those instead. You need to use a permanent marker to write on the card, but the information stays legible longer.

Preparing to Harvest

Generally, heritage breeds take a bit longer to mature and reach processing age than commercial stock. Twelve to fifteen weeks of age should be a reasonable target, but the best indicator is weight. Five to six pounds (2.3–2.7 kg) seems to be a good target weight: much bigger and your feed conversion ratio goes down; much smaller and the meat yield is not nearly as good. Keeping track of litter weights is valuable; rabbits that reach processing weight earlier than their littermates might be selected as replacement stock.

Most heritage breeds, and most breeds overall, will dress out at 50–55 percent liveweight. This seems to be a consistent dress out percentage across the board. You will occasionally have animals dress out slightly better, and sometimes slightly worse, but this number provides a good rule of thumb.

CHAPTER 13

Processing, Storage and Distribution

We in North America live in a generation and are part of a world that has lost touch with the basic skills and necessity of harvesting our own meat for the table. Our parents grew up with parents and grandparents who, as part of their daily lives, butchered their own meat and had a healthy respect for the process. Many a rooster crowing on Saturday was the guest of honor for Sunday dinner.

When so many North Americans left farms in the 1940s and '50s, we stopped providing our own food directly. Now, meat is a shrink-wrapped convenience product. Consumers of that meat have little to no connection to what provided it or what it truly cost.

The fact that you are reading *Raising Rabbits for Meat* and have made it this far means you are different and that is amazing and encouraging. So now we come to the pinnacle of our efforts.

Stopping to Consider

You have been working towards this goal for the last few months. Time to put the rabbits into the freezer. Right?

It's normal at this moment in time to feel some trepidation. Harvesting an animal you have raised from birth is not something to take lightly.

If killing and processing a rabbit doesn't bother you in the least, that doesn't make you a bad person. It doesn't mean you're weak if it does. The important thing is to be honest with yourself.

Processing Equipment Checklist

☐ **Broomstick** or **rabbit wringer** for stunning

☐ **Sharp knife** (Invest in a good butchering knife. This is not the time to use that old steak knife clattering around in the kitchen drawer. A good boning knife is also essential.)

☐ **Hooks to hang the carcass on** (Skinning hooks are available from most rabbit supply places. In a pinch, a heavy-duty staple can be bent to hang the rabbit from.)

☐ **Bucket for offal** (Inedible parts can be composted if you do not have a problem with predators or racoons digging things up.)

☐ **A plastic butchers apron**, or the like (You need something to keep your clothes from contaminating your meat, and to keep inedible parts from contaminating your clothes. You should rinse this apron periodically through the process, helping to keep the area clean.)

☐ **Rubber boots**

☐ **Nitrile gloves** (If you have recent cuts on your hands, or if you are just learning knife skills, a Kevlar glove might be a good investment. If you keep your knives sharp (as you should), it can be very easy to nick or cut yourself, especially as you are learning. (There's a very good reason Eric does all the knife work in our house…I once cut myself badly unpacking a food processor.)

☐ **Food-grade tub** (This does not have to be some special expensive item, a clean plastic storage container will do. However, it does need to be rinsed and bleached and as clean and sanitary as you can make it. Food you are going to eat later is going into this, so cleanliness is vital!)

☐ **Ice water** (An ice water bath begins chilling the fryer, reducing the chance that bacteria will begin to grow on the meat.)

☐ **Garden hose/water source for rinsing the carcass** (This will be important for keeping the carcass clean, removing hair, and keeping foreign material from the carcass.)

☐ **Cutters of some sort** (Almost any good cutter will do, e.g., a pair of loppers, an anvil cutter. While you can use your knife to remove feet, it is hard on a knife to cut through that much bone and tendon, and if it slips, you may be headed for an emergency room visit. You basically need something stout enough to cut tough pieces without being awkward and unhandy.)

Once you have reflected, and made your decision, the time for hesitancy is past. Hesitancy can ruin the process, and cause unnecessary pain and suffering to your animals, and yourself.

If, when it comes down to it, you feel like you are going to struggle with processing the rabbits, ask for help. Find an experienced rabbit farmer—or a hunter with experience in field dressing small game—to show you how to process the meat the first couple of times.

Meat requires sacrifice. All life does at some point or another.

Stunning

One of the first things you will need to figure out is the method of stunning. It is unethical to process an animal, any animal, without rendering it insensate before starting.

For rabbits, this can be accomplished in two ways: cervical dislocation and blunt force stunning.

Shooting is not recommended. Why? For one, it's overkill. Pun intended. If you miss, you can destroy a bunch of your meat, and if you are not careful where you set up the kill shot, it can be dangerous.

Stunning causes such massive disruption to the system and the neurological activity of the brain that the rabbit is unable to process stimuli. It's important to work quickly from that point on, so if you feel it's something you will be hesitant about the first few times, practice the moves before bringing a live rabbit into the equation.

Blunt Force Stunning

Use an end of a broomstick, ax handle, or something else short and stout, that can be swung quickly and forcefully.

Hold the rabbit by the hind legs and stretch the legs up and backwards to lengthen the rabbit's body and expose the back of the neck.

A sharp, swift blow to the back of the head will render the rabbit unconscious. This is not a time to be hesitant.

Blunt force stunning can make you feel a little brutal, even though when done correctly it is a quick, efficient process.

Cervical Dislocation

Cervical dislocation requires a little more finesse than blunt force stunning, but it's much less likely to go wrong.

The *broomstick method* of dislocation involves a broomstick or another long piece of wood, metal or concrete rebar. The rabbit's neck is placed under the stick, and the person steps on either end of it, while still hanging on to the rabbit's hind quarters. The hindquarters are then pulled up sharply, dislocating the neck at the point of pressure. The major drawback of this is that if you don't have great balance, or the rabbit is very large, it can be stressful for the rabbit while you get the bar across its neck, step on both ends and get yourself into position.

In our experience, a much more humane method of cervical dislocation uses a device called a *rabbit wringer*. These can be purchased,[1] or if you, or someone you know, is handy with a welder, they can be created relatively easily. The wringer is attached to a solid wall, table, or tree. The rabbit's neck is slipped into the V, and with a sharp downward pressure, the neck dislocates easily. Done well, the rabbit is killed in a second or two and never has a chance to get anxious. It works well on any size of rabbit, and for a smaller person it's much easier to just focus on manipulating the rabbit quickly and exactly.

In order of preference, we recommend the wringer, stunning, and lastly the broomstick method. We've seen things go wrong with both of the latter two; so far the wringer has never let us down.

If it goes wrong? I feel like at this point there should be a word of warning about rabbit vocalizations. A rabbit scream (because there is really no other word for it) is a really awful and heartbreaking sound, especially if you've never heard it before. Rabbits can scream for no apparent reason—simply because you pick them up, looked at them funny, or because it's Tuesday, but

FIGURE 13.1. Homemade rabbit wringer.

if stunning or broomsticking goes wrong, you may get to hear it.

Don't panic. Regroup quickly, and try again. And don't beat yourself up too much. Experience requires practice. You will do better next time.

Hanging

After the rabbit is stunned, working quickly, take the knife and put a small holes in the leg, right above the hock joint. There is a place in between the leg bone and the tendons, that works quite well to hang the rabbit from.

Place the rabbit on the skinning hook by slipping the leg where you put that hole over the hook. You will need a sturdy place to hang the rabbits from, as a fair bit of force is generated in skinning them.

Once the rabbit is hung, remove the head. Take your knife and cut between the back of the head and the shoulders, slicing downward and out, through the neck. (You may or may not want to discard the heads as there are a couple of products that can be made from them.)

There will be a surprisingly small amount of blood. Rabbits do not have a huge blood supply—only 6 ounces or about ¾ cup. A rabbit carcass will only take a couple of seconds to bleed out.

Rinse the carcass while the skin is still on. This keeps dirt and blood to a minimum once it is skinned, and also keeps hair from flying around quite as much. The wet hair will stick to the pelt better.

FIGURE 13.2. Rabbit hanging before skinning.

FIGURE 13.3. Gambrel hooks support a rabbit carcass.

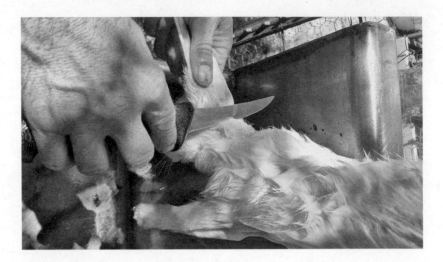

FIGURE 13.4. Removing the rabbit's head.

Skinning

While the rabbit is hanging, you can loosen the skin at the point where the gambrel hooks go in, pulling it loose and pulling it down the legs.

Make a small cut in the skin between the hind legs. Use this cut to stick your fingers in and begin to loosen and stretch the hide apart from the fascia. Begin working the skin down over the body.

FIGURE 13.5. Pulling off the rabbit's skin.

Rabbits are a relatively easy animal to skin. The hide should pull down relatively easily. It does become much tougher with older animals, but young rabbits at about 8–12 weeks at processing time will skin fairly easily.

Now, keeping the meat cool and clean becomes critical. Rinse the skinned carcass thoroughly.

Once the hide is removed, it will be attached still at the front paws; take a pair of cutters and remove the front paws and the pelt. Set these aside.

Now make a small incision in the belly. Slice downward, with the blade of the knife pointing outward to avoid nicking intestines or the bladder.

Reach into the cavity, and pull the organs out.

The liver, heart and kidneys are all edible. The stomach, intestines and lungs are not. The edible organs can be placed in an ice water bath.

Discard inedible viscera. Rinse the cavity thoroughly.

Remove the back feet with your cutters. You should now have a completely dressed rabbit. Place the whole rabbit in ice water to chill.

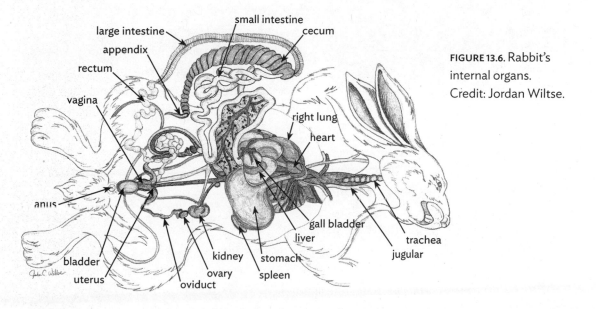

FIGURE 13.6. Rabbit's internal organs. Credit: Jordan Wiltse.

FIGURE 13.7. Opening the cavity.

If it's helpful, write out the steps on a note card and pin it right near where you are working.

You may notice how easily debris and foreign material stick to the fascia surrounding the rabbit fryer, compared to other meats you may have processed. This debris and dirt will be terribly difficult to rinse off, so be super diligent about keeping equipment, work surfaces and your own hands clean.

Storage

The whole dressing process, start to finish, should only take a few minutes. An experienced rabbit processor can probably take a rabbit from start to fryer in two or three minutes. It will likely take you a bit longer the first few times you do it, but with practice and familiarity, processing will become much easier and efficient.

Once you have your fryers processed, they will need to be chilled. Chill them overnight, and then prepare them to go in the freezer the next day.

Put the carcasses in plastic bags so they do not begin to dry out. The meat is fairly delicate and can dry out quickly. It's fine to put more than one carcass in a bag to chill, but it's best to bag them individually for freezing.

Do not leave them in the ice water and put them in the freezer or the refrigerator. After a certain period of time, even if chilled, the water becomes not so clean, so discard the water before either chilling or freezing.

Chilling and Aging

After death, the complex process of enzyme activity causes a mammal's muscles to move from a state where they can contract and relax into a state of total contraction (known to us as *rigor mortis*). Aging meat is

FIGURE 13.8. A whole rabbit carcass chills in an ice water bath.

a process in which these naturally occurring enzymes then work on the muscle proteins, breaking the bonds that cause the contraction. For larger animals like sheep, hogs, or beef, aging takes several days to weeks, but in rabbits it only takes a day or two.[2]

Putting the rabbits after processing and before packaging into the refrigerator will allow them to age, but keep them at a temperature which will deter bacterial activity.

Packaging

If you plan to cook your rabbits very soon, within a day or so, vacuum packaging isn't necessary. But to store them in the freezer for any length of time you will need to vacuum package them in some way.

One of the most common ways to do this is to use one of the readily available *home vacuum packaging* systems. We've tried a couple of them, and they all work pretty much the same. These gadgets use a custom plastic bag that, when placed into the machine, draws air out and heat-seals the bag closed. The plastic can either come in standard sizes, or in rolls that you can cut to your specifications. The rolls work quite well for rabbits, which are often a longer and more oddly shaped objects than the custom bags are designed to hold.

Another method is to *shrink-wrap* them, putting the rabbits in specific shrink-wrap bags and dipping the bagged rabbits into hot water.

Air in the package is the enemy of long-term storage. Air in the package allows ice crystals to form, which pull moisture out of the meat, resulting in a condition known as *freezer burn*. Oxymoron aside, freezer burnt meat, while still safe to eat, is unpalatable and tasteless and a profound waste of an animal.

Hot water causes the shrink-wrap plastic to constrict around the rabbit, preventing contact with air. The challenge with this system (with either system really) is to get all the air out and maintain a good seal around the meat.

When packaging, be careful of sharp bone ends on the ends of the feet. These can poke even a tiny hole in the bag and make the vacuum package useless. One way around this is to tuck the hind legs into the chest cavity, and fold the front ones up so they do not stick straight out.

Whichever method you choose, monitor the conditions in your freezer and watch your packaged meat for any signs that the seal around the rabbits has become loose. Any that have should be eaten right away to avoid wasting the meat.

Food Safety while Processing

Cleanliness

Cleanliness is indeed next to godliness when it comes to processing meat at home. The home processor has an advantage simply because they are not exposing their animal to every other animal that has passed through a big facility, but that doesn't mean sloppiness is appropriate or encouraged when butchering at home.

One of the simplest and most effective things you can do to prevent contaminating the meat is to *wash your hands*, well and often. Spend the extra few seconds it takes to sing Happy Birthday twice, count to 30 while you are scrubbing your hands, or watch a clock; food safety experts recommend washing your hands for 30 seconds under running water that is as warm as you can comfortably stand.[3] This washing will mechanically remove any germs from the surface of your skin, and allow them to be carried away with the water. Washing your hands with soap won't as much kill pathogens as it will physically remove them. Dry your hands with a paper towel, or clean towel.

Take the time to *clean the surfaces* you will be working on with a weak bleach solution, and allow those surfaces to air dry. Constantly clean the surfaces while you are working on them. Don't allow scraps of meat to accumulate on cutting surfaces when breaking the rabbit down, and likewise keep everything rinsed in the processing area.

Temperature

The simple fact is that we are never going to be working in a completely sterile environment no matter what steps we take. Germs happen. And, ultimately, bacteria will have their way with us all.

However, we can take firm and confident steps to reduce the amount of bacteria, and severely limit their ability to reproduce. The so-called *danger zone* for food safety is between 40°F and 140°F (4°C and 60°C). This is the temperature range at which bacterial activity can really ramp up.[4] And, the higher the temperature gets, the faster bacterial activity becomes. Keeping meat cold does not eliminate bacterial activity, it merely slows it down to allow us to consume it safely. Food will eventually spoil even in the coldest refrigerator. Food safety experts estimate that food has about four hours total in that danger zone before it becomes compromised.[5] And that's four hours total, not four hours at each stage, so keep that rule of thumb in mind as you work with the meat through the cooking process, and through handling leftovers.

Another caveat is most refrigerators have inconsistent temperature zones. Even if you place a thermometer in one spot and get a satisfactory reading, the temperature may not be the same in the rest of the unit. The only way to be sure is to take multiple readings, and adjust temperature setting accordingly. Also, placing a large quantity of warm meat in a refrigerator will increase the temperature inside that refrigerator; it will take some time to get back to the desired level. Factor that in when setting your fridge temperature, and also remember

> You may have noted that human body temperature (98.6°F [37°C]) falls smack in the middle of that temperature danger zone. This is why foodborne illness can come on so quickly; our bodies are literally at an ideal temperature for bacteria to reproduce.

that if you put several rabbit carcasses in close together, it will take more time for the inside ones to reach a safe temperature. Try, as much as possible, to allow air to flow around each package to speed the chilling process.

Cooking

The good news is that bacteria and germs generally only contaminate the surface of meat—until you puncture or grind it. This is also why many, many of the food recall notices in the US involve ground beef.[6] Mixing product from multiple animals or increasing the surface area exposed to potential pathogens can make meat a literal breeding ground for bacteria. The external surface will reach a temperature that kills most bacteria, or at least reduces it to a level that a healthy immune system can handle with no problem. We will go into a little more detail about cooking in the next chapter.

Retail Sales

In the US, United States Department of Agriculture (USDA) inspection is required for sales of meat to restaurants or shipping across state line. Period. There is just no way around this law, and reputable chefs won't risk their operations by trying to circumvent it.

Each state, however, has different guidelines for in-state sales, and if you want to sell rabbit meat to others, it would pay you to spend some time researching what your state allows. Some states will allow several hundred rabbits to be sold at farmers' markets, others none at all, and each state differs in what they will allow to be part of *farm gate sales*. Laws are constantly changing and being updated, so before diving in to a retail sales operation, familiarize yourself with what you can and can't do where you live. There are always options, but don't try to circumvent the law. Evading regulations can only bite you in the end.

Our business is based on USDA-inspected processing of our rabbits. This has probably caused the most headache and challenges for us during the course of our operation.

USDA inspection of rabbit is voluntary which, in my opinion, is a terribly inaccurate word to describe the fact that inspection costs are borne completely by the farmer. No subsidized inspection, period. Inspection costs ultimately come out of our pocket, and it isn't cheap. If you are used to having beef or other animals processed at a local facility, the additional cost for rabbit inspection can give you a bit of sticker shock.

USDA processors that are willing to work with rabbit are few and far between, at least in the Midwest. I believe there are more opportunities on the east coast, but it will be rare that you find one very close by. Be sure to take that into account inspection costs when planning your operation.

Shipping

Shipping perishable goods so they arrive in a safe and usable form is also an art that we had to learn.

Shipping boxes must be durable: heavy-duty cardboard on the outside, and insulated on the inside. There are several companies that make Styrofoam shipping containers, but these are not cheap. We've wound up creating our own with heavy-duty boxes and foam board cut into squares to make a liner for the box. The best quality, most reasonably priced boxes we've found have come from our local large chain hardware store. These are generic, heavy moving boxes, and they work well for shipping 12–14 rabbits, depending on weight. Carriers charge extra if a box weighs more than 70 pounds (32 kg), and you will need to take into account the weight of the box, the meat, and ice packs you will be enclosing.

We do put perishable stickers on the boxes, but no other information about the contents. A good friend told me a story years ago about one of the major carriers having an employee that had some personal issues with meat consumption and any boxes that were clearly labeled as such were kicked to the side, and allowed to spoil. That employee no longer has a job, but that story stuck with me pretty profoundly.

Harvesting meat and consuming it is one thing; deliberately causing it to go to waste is simply unethical and fires me up to a level no one wants to see.

Unless the product you are shipping is frozen solid, you have about 24 hours to get it from point A to point B.

Most carriers have a three-day ground rate that is relatively reasonable, and we have been lucky in that most of our shipments are close enough to our operation that they will arrive the next day, even with sending them the three-day rate.

This has been relatively problem-free (knock wood), but the potential for spoilage always exists. Carriers will not guarantee the package will arrive on time; that's simply their best estimate based on route times. And, you have no recourse if it does not.

Here are a few tips for minimizing problems shipping with a major carrier. And yes, these are also ones we've learned the hard way.

- Ship early in the week. Even if your shipments regularly take one day, shipping on Thursday, especially during busy shipping seasons, means that if the driver can't complete the route on Friday, your package will sit in a truck or a warehouse until Monday.
- Don't be stingy with ice packs, especially in the summer. And consider shipping frozen in hot weather. Restaurants will reject a shipment if it arrives at a suspect temperature, and that loss will be yours.
- Make sure your customers know when to expect delivery. After a decade, our customers know pretty much to the hour when their order will arrive, but I still send a text or email just as a reminder.

The majority of our business is also not in our zip code. This necessitates using a major carrier to get our rabbits from the processor to the destination. We have been fortunate in that most shipments can arrive in that one-day window, allowing us to ship most of our orders fresh. We looked into delivering, but by the time the costs got penciled out and our time was factored in, it made more sense to let the carriers transport our meat for us.

CHAPTER 14

Cooking with Rabbit Meat

Nutritional Information

Rabbit is one of the healthiest meats you can consume. Naturally low in fat, it is high in protein, iron and B vitamins, among other good things.

According to the USDA National Nutrient Database, a site which literally contains thousands of reports of nutrient information, rabbit ranks higher than beef in a couple of desirable categories.[1]

For simplicity's sake, I chose a beef tenderloin for comparison, since that is one of the most desirable beef cuts, but you could spend all day looking through and comparing dozens of different cuts, and values also differ, depending on how meat is cooked—for example, how much fat trim is on the cut.

There is much less information about rabbit, and the comparisons are of whole rabbit, and a composite of cuts.

Now we all know by now, fat (good fat, let's be clear) is not the dietary demon it was made to be during the last few decades. But not all beef is created equal, and the benefits of beef depend on that meat being the right cut, from the right animal, that was fed the right diet.

Rabbit will have a more consistent nutritional content, making it a more reliable and healthy choice.

Rabbit is surprisingly high in iron for a white meat, and also higher in some of the B vitamins than one might expect. It is overall lower in calorie content (energy kcals) but higher in protein overall, and definitely higher in protein per calorie. Combined with the low content of fatty acids, rabbit is an ideal meat for the weight conscious, and those who want to maximize their caloric value.

Cooking Rabbit

Part of the reason for rabbit's leanness is that it has little to no intramuscular fat. Unlike other animals, even a rabbit fed to excess will never put on a *fat cap* (a layer of fat under the skin), and rabbit meat will never marble. Rabbits lay down fat in a narrow cross over the shoulder blades, and in the abdomen, but not in muscle.

This lack of fat can present a challenge for cooks who are not familiar with cooking lean meat. Fat is very forgiving of high temperatures, and it can take a little practice to get comfortable cooking with rabbit, but the benefits are worth it. The key to remember: low and slow.

TABLE 14.1. This comparison of rabbit and beef tenderloin nutritional values is based on a 100-gram (around 3½-ounce) portion.

	Rabbit	Beef
Energy (kcals)	197	331
Protein (g)	29.06	23.9
Fat (g)	8.05	25.39
Calcium (mg)	19	9
Iron (mg)	2.27	3.11
Magnesium (mg)	21	26
Phosphorous (mg)	263	236
Potassium (mg)	383	407
Sodium (mg)	47	65
Zinc (mg)	2.27	4.03
Thiamine (mg)	0.09	0.15
Riboflavin (mg)	0.21	0.27
Niacin (mg)	8.34	3.88
Vitamin B_6 (mg)	0.47	0.48
Folate (µg)	11	8
Vitamin B_{12} (µg)	8.3	3.37
Cholesterol (mg)	82	85
Saturated Fatty Acids (g)	2.4	10.2
Monounsaturated Fatty Acids (g)	2.17	10.5
Polyunsaturated Fatty Acids (g)	1.56	0.96

Braising

Rabbit does very well in recipes that call for braising. Braising is a fancy word that simply means browning meat in oil, and then cooking it in a covered pan with a small amount of liquid, at a low temperature for a comparatively longer time.

A cast-iron Dutch oven is ideal for this. An oven temperature of 225°F (107°C) is great, and the length of time? Each oven is different, and it depends on whether you put the rabbit in whole or not.

A whole rabbit will cook a little less evenly than if it is broken down. And the meatier hind legs cook more slowly than the leaner, smaller front legs which can dry out if not watched carefully. In our house we usually just eat front legs when we check the rest of the portions for doneness, and call them appetizers.

The USDA recommends cooking rabbit to 160°F (71°C) for food safety.[2] I'll be honest, I don't even cook my beef to that sort of temperature, and I have minimal concern about any potential risks with rabbit we raise and process at home. But that's the recommendation, and I'd be remiss if I didn't mention it.

Rotisserie

This is hands down Eric's favorite way to cook rabbit. Our outdoor grill has a rotisserie attachment, and since the grill is his domain, rotisserie cooking has become my favorite way as well!

We mix equal amounts of melted butter and zesty Italian dressing in a cup. Using one of the meat injection syringes we got at our local supermarket, Eric injects the whole rabbit in several places with the mix, and puts it on the rotisserie skewer. (Tying the rabbit with butcher's string or gauze keeps it from slipping off as the meat shrinks and cooks.) On the grill for about an hour, basting with the butter/dressing mixture regularly, the meat is super tender and delicious, and many times we just eat it straight from the grill.

Crock-Pot

The Crock-Pot is a good friend to rabbit meat. A Crock-Pot can cook up a tasty batch of rabbit noodle soup or a stew with very little effort. The rabbit can either be cut apart or left whole, and covered with chicken broth (or better yet, rabbit broth), and left for several hours to stew.

Depending on your goal for the rabbit, after cooking for several hours the rabbit can be shredded for soup, tacos, BBQ rabbit, whatever your heart desires; or the portions can be served whole, with a sauce made from the cooking broth.

Adapting Chicken Recipes

Basically any chicken recipe can be adapted for rabbit. Rabbit pieces are largely the same, can be fried, or roasted whole or whatever your heart desires. Just keep in mind that chicken will have a skin to keep moisture in, and rabbit will not.

Recipes

Our story is forever linked to the restaurants who have faithfully used our rabbits over the years. Many of them have done some mind-boggling things with rabbit, showcasing both the chefs' talents and creativity and rabbit's versatility. When we first started talking about doing this part of the book, we knew that we wanted to include contributions by those chefs if at all possible.

We could not have done what we have done without them. They are always quick to praise the quality of our rabbit, but after all, you could give a chimp a Steinway, and you wouldn't get beautiful music. And the great thing about all of these recipes is that they are not complicated or fussy. A couple may look intimidating at first because they are expertly seasoned, but they all rely on basic cooking techniques and are well worth the effort.

Clay Pot Rabbit

from Chef Michael Foust at The Farmhouse

1 rabbit

2 carrots (medium cut)

1 parsnip (medium cut)

1 sweet potato (medium cut)

4 cloves garlic sliced thin

2 leeks (clean, medium cut, whites only)

½ pie pumpkin (medium cut, peel, seed [save seeds])

1 fennel bulb (medium cut)

1½ cup red wine (Pinot)

16 shiitake mushrooms

4 cup rabbit stock

1 bay leaf

3 sprigs of fresh thyme

1 tbsp. each salt and pepper

¼ cup flour to coat rabbit (coat and brown)

¼ cup sunflower oil

powdered sumac

½ tbsp. sorghum

¼ cinnamon stick

1. In Dutch oven or clay pot with lid, heat the oil on medium-high heat. Dust the rabbit with flour and brown. Remove rabbit. Add vegetables, cinnamon, bay leaf and thyme wrapped in cheesecloth.

2. Cook 5 minutes and deglaze with wine. Reduce by ⅓. Add stock. Bring to boil, cut to simmer. Add sumac and sorgum. Season with salt and pepper to taste.

3. Put rabbit back into pot and cover. Liquid should go ⅓ the way up the rabbit.

4. Cook at 350°F (177°C) for 2–2½ hours. Meat should fall off bone with a little pull.

5. Serve with rustic farm bread.

Smoked Rabbit Sausage   *from Chef Vaughn Good*

Yields about 20 sausages.

3 lb rabbit, deboned

1 lb cubed pork shoulder

1 lb cubed bacon

Dry

5 grams Instacure #1

2 tbsp. Kosher salt

1 tsp. dry sage

17 grams minced garlic

2 tsp. dry thyme

12 grams black pepper

1 tsp. ground allspice

1 tsp. mustard powder

Wet

¼ cup cider vinegar

½ cup water

pork middle natural casing

1. Combine cubed meat with all dry spice and minced garlic.
2. Grind once through a large grinder die and then once through a small die.
3. Combine wet ingredients and mix into ground meat.
4. Cook off a small amount and check for salt.
5. Using a sausage stuffer, stuff into pork middle natural casings. Link in about 4-ounce links.
6. Smoke at 150°F (66°C) with a 50/50 combination of hickory and cherrywood. Smoke to an internal temperature of 140°F (60°C).

Braised Rabbit

from Chef Michael Beard

1 rabbit cut into pieces
mirepoix large cut (1 onion, 2 carrots, two stalks celery)
2 cloves garlic, smashed
1 qt. chicken stock
¼ cup tomato paste
1 cup red wine
bunch thyme
2 bay leaves
¼ cup oil (high temperature oil, e.g., grapeseed, peanut, avocado)

Season rabbit legs liberally with salt and pepper. Let sit overnight in cooler (or at least for a couple of hours). Take rabbit out of the cooler and let meat come to room temperature. Heat large rondeau pan with oil until almost smoking and sear rabbit pieces until golden brown. After searing, turn down heat, add mirepoix and carmelize. Make a well in the center and add tomato paste and cook until paste carmalizes a little and then mix with vegetables. Add wine and reduce until there is no wine smell. Add stock and bring to a simmer. Place legs in a baking dish and add stock mixture. Cover with foil and bake in a 210°F (99°C) oven for 3 hours. Remove legs and strain cooking liquid over legs and let cool. Label and date. Best if left overnight to marry flavors.

Rabbit Posole   *from Chef Lee Meisel*

Ingredients

1 whole rabbit

1 27-oz can of crushed tomatillos

1 27-oz can of diced hot green chilies

1 small can of tomato paste

2 27-oz cans of hominy, drained

1 large onion

3 jalapeños, finely chopped

3 cloves of garlic, minced

2–4 tbsp. New Mexico chili powder

2 tbsp. ground cumin

1 tbsp. ground coriander

1–2 tsp. cayenne pepper

2 tsp. Mexican oregano

salt and black pepper

oil

1 12-oz can Mexican beer

Garnishes

lime wedges

diced raw onion

sliced radishes

cilantro leaves

Meat

Place the whole rabbit in a large stock pot, cover with water and place a parchment lid on top to keep the meat submerged. Bring up to a simmer (don't boil) and poach the rabbit until the meat is cooked through but still firm on the bone. Turn the heat off and let the rabbit cool in the poaching liquid.

Soup Base

When it is cool enough to be handled, pick the meat off the carcass and place in a bowl. Take the remaining bones and return them to the pot along with the poaching liquid and simmer for one hour. Top off water as necessary to keep the bones submerged. Strain the liquid and reserve. You may now dispose of the bones.

Aromatics

In the same stock pot add oil and sauté the onions, jalapeños, and garlic over medium heat until softened. Add the cumin, coriander, chili powder and oregano and cook for 30 seconds, stirring constantly to prevent burning and adding more oil if necessary. Add the can of tomato paste and cook for another minute or until the tomato paste has lost its bright red color. Deglaze with the beer, carefully scraping any food from the bottom of the pot. Add the canned chilies and tomatillos and cook on medium-high heat for another 5–10 minutes, stirring occasionally to prevent scorching.

Finish and Plating

Add the stock and hominy to the pot of aromatics and bring to a simmer. Avoid boiling. Season to taste with salt and pepper. Cook for 15 minutes and then add the picked rabbit meat. Simmer for another 15 minutes. Check seasoning one last time.

To serve, ladle soup into bowls and garnish with diced onion, radish slices, limes wedges and cilantro leaves, and your favorite hot sauce.

CHAPTER 15

Rabbit Ailments and Health Problems

Rabbits are relatively hardy animals, and good husbandry can go a long way to keeping stock healthy and productive.[1]

There are a handful of problems that are most common in the rabbitry, a few of which are relatively minor and easily treated, and a few which are more serious. Veterinary attention for rabbits can be hit or miss; many vets are not familiar with rabbit medicine, and treatments can be expensive, depending on the problem. And in the case of meat rabbits, treatment may not be worth the economic cost.

Do not feel bad for culling an ill or injured animal. Do feel bad for letting an animal suffer needlessly.

Like most prey species, rabbits can be quite stoic when ill. Lying around moping when not feeling well is a sure invitation for a predator to grab a quick lunch, so rabbits tend to keep going, and domestic rabbits still share that stoic attitude with their wild cousins. Serious illness can seem to progress quite rapidly in rabbits, and in some cases, the first sign of sickness is a rabbit that is very nearly dead. Some of the conditions described in this chapter you may fortunately never encounter; some you might be challenged by on a regular basis. While it's not intended to be an exhaustive list, what follows will hopefully give you some basics to build on.

Snuffles

Snuffles is the rather benign-sounding name given to problems resulting from the bacterium *Pasturella multocida*. It is one of the most

FIGURE 15.1. Snuffles dampens the paws and arms of an affected rabbit. Be sure when looking at paws that the rabbit has not recently eaten or drank; Eric usually inspects animals before watering just to make sure that isn't a factor.

prevalent problems in the rabbitry. Some studies report that *Pasturella* is present in 75–80 percent of rabbitries, and even research facilities are not immune.[2]

The bacteria may be present, but remain dormant, and not show any disease symptoms until the rabbit is stressed by change in environment, the challenges of reproduction, or its overall health becomes compromised, either by poor air quality or some other environmental factor.

Chronic snuffles is categorized as an upper respiratory tract infection, and exhibits symptoms such as a white nasal discharge, sneezing, and often matted fur on the front paws, a result of rabbits trying to clean its face and nose.

Animals with snuffles often begin to lose condition, and their growth rate can be compromised. They can also have a generally unthrifty appearance. Pneumonia is often a secondary complication.

Treatments for snuffles are often unsatisfactory at best. Antibiotics can be used, but often just reduce the infection to a subclinical level where it can reoccur during another stressful period. Treatment may help with the secondary infections, but the snuffles will always be present.

The only sure way to remove the problem is to cull the infected rabbit. This sounds ruthless, but snuffles can cause significant complications, and it can be spread to an entire herd by sneezing or by the bacteria being carried on drinkers or shared equipment.

Strict culling can ultimately result in rabbits that are at least somewhat resistant to the clinical infection. If your best breeding stock is among the animals infected, it is possible to remove the young at weaning, place them in a separate, well-disinfected pen that is some distance from the infected animal, and then watch the litter intently for any symptoms, removing any symptomatic animals as soon as you discover them. There are no guarantees, but this might allow you to discover and retain animals that have the better resistance to this infection.

If a doe is sneezing heavily, chances are her litter has already been exposed, and the odds of finding animals not already heavily challenged by the bacteria are slim.

Good ventilation and sanitation in the rabbit barn will go a long way to keeping snuffles from becoming a major problem and limiting economic losses. Humid air heavy with ammonia fumes is an almost sure way to give snuffles a foothold in your rabbitry.

Abscesses

Abscesses in the rabbit are often the result of *Pasturella* bacteria as well, but treatment is much more successful. The abscess can be lanced with a scalpel blade and drained, and flushed with a weak solution of iodine, chlorhexidine or peroxide.

A draining abscess will often produce a white to yellowish material, ranging in consistency from liquid to the relative consistency of toothpaste. Be sure to flush until only a little of this material can be extracted.

Recovery and prognosis are generally excellent. Be sure to use good sanitation when draining abscesses: wear single-use gloves to avoid carrying the bacteria around, and clean all the tools you use with a bleach solution and a little strong sunlight.

If an abscess happens to pop and get goo all over the rabbit pen, scrub everything clean as soon as possible, and spray a dilute bleach solution in the pen. And be sure to wash your hands thoroughly after handling a rabbit with an abscess, as well as the equipment used for cleanup.

Abscesses seem to be polarizing: some people find the discharge disgusting, others find it fascinating and are eager to lance and drain an abscess when needed. Hopefully, if you are in the first camp, you can find someone in the second to help you deal with them.

Ear Mites

Ear mites are one of the most common and most annoying rabbit parasites. Fortunately they are also one of the easiest to cure.

Ear mites are *Psoroptes cuniculi*, a nearly invisible little critter that lives on the surface of the ear and in the ear canal. They bite and feed on ear secretions and oils, as well as skin flakes, irritating the ear and causing excessive wax production. An affected rabbit will often be scratching its ears, and shaking its head. Scratching can also lead to further infection in the ear. Rabbits with a severe infestation of mites can be miserable and restless, and go off feed. Their ability to thermoregulate via blood flow in their ears also seems to be compromised, which can lead to heat stress, and a reduced ability to resist heat.

Treatment is simple and effective. A couple of drops of mineral oil in an affected ear will smother the mites, causing them to die. The rabbit will then clean the debris out of its own ears. To treat effectively, apply mineral oil three days in a row. A maintenance dose can also be given once a month. In our rabbitry, each buck gets a dose of mineral oil on the first of the month. The does get their ears treated whenever they are bred, or right after kindling, basically every time they are weighed.

Some rabbit people have used Ivermectin as a treatment, but we have never found that to be necessary. If a rabbit fails to be cured of

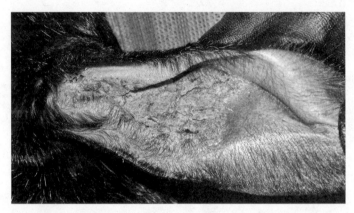

FIGURE 15.2. This heavy mite infestation was only allowed to get this bad for the photo!

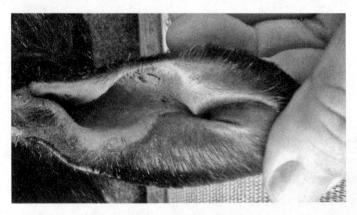

FIGURE 15.3. Infected rabbit ear two days later.

an ear mite problem, even after successive rounds of mineral oil treatment, consider culling that animal.

Heat Stress

Heat stress, while technically not an ailment is one of the most common problems affecting the health and well-being of rabbits, and it's the one that can have the most impact on your bottom line.

Rabbits do not sweat. Their highly vascularized ears cool them by acting as a radiators; and rabbits also pant, which exchanges warm air from the lungs for cooler air outside the body.

On days when the temperature is extreme, rabbits may not be able to exchange enough body heat to adequately reduce their core temperature, and may experience severe heat stress.

When it is hot, rabbits will spread out in their pen, and begin to pant. This is normal. They will also appear to be "zoned out." Make every attempt to leave rabbits alone during hot weather, and absolutely avoid any procedures such as tattooing, breeding, transporting, or anything that can cause the rabbit any discomfort or stress.

An animal experiencing heat stress is at serious risk of death. Signs include open mouth breathing, and dampness around mouth and nose from respiratory condensation. This condition can be fatal. If you find a rabbit in this condition, it can be cooled by submerging it in a bucket of cool—*not cold*—water to bring its core temperature down. Excessively cold water can cause shock and can easily bring about a stroke or death, exactly what you are trying to prevent.

A fan can be set up for the rabbit to move air around it, but if the air is not cooler than the ambient temperature, this may not prove effective.

Many rabbit barns keep water bottles in the freezer and put them in with the rabbits on hot days. This is truly only effective if air can move over the rabbit, and the rabbit leaves the ice water bottle alone and does not move it to a corner of its pen. Considerable freezer space can be tied up with water bottles, and if you are not there to make sure the bottle is frozen during the hot part of the day, it will have little effect.

Frozen bottles are not something we have ever made a practice of using, both because the sheer number of ice bottles it would take for each rabbit pen would require the use of an industrial-sized freezer, and a small army to keep them changed out.

Over the years, we have been able to select animals that seem to have the best ability to withstand the intense Kansas summer heat, and we have very little problem these days loosing rabbits to heat stress. Eric is also adamant about allowing the rabbits relax and "zone out" during the hottest part of the day. Misters set up around the barns (see Chapter 7) also help keep temperatures comfortable.

Avoid bringing hot animals into the house or into air-conditioning. While this initially seems like a good idea, household air might not be the best for the rabbit, and the changing environment can be stressful, especially if you have house pets that will be fascinated by the rabbits.

Malocclusion (Bad Teeth)

Malocclusion in rabbits usually refers to the front incisors. In a normal mouth, the top and bottom incisors line up, with the lower ones sliding slightly behind the top ones. This gives the rabbit the ability to shear off bites of food, whether that is pellets, grass, hay or another food item. Rabbit teeth continually grow throughout their lifetime, and depend on proper alignment to keep teeth worn down and in line.

If the teeth do not line up properly, they will not wear evenly, and you will begin to see one or more teeth grow at strange angles. This will ultimately make it hard for the rabbit to eat, cause serious loss of condition, and ultimately could lead to the animal starving.

Malocclusion is usually hereditary. Animals that have poor teeth should not be selected for

FIGURE 15.4. Misaligned and crooked teeth.

breeding, although they can still be fine for the freezer. Severely oc-cluded teeth can be trimmed with tooth nippers of the kind used to clip needle teeth in baby pigs. Be careful that what you do use does not just mash the tooth and shatter it down to the root.

Malocclusion can also be caused by injury, when a rabbit bites the feeder or pulls on pen wire. If the animal is a promising individual that you had hoped to select as a herd animal, consider the factors below to determine whether malocclusion is genetic or environmental.

> ► If there are other litters from this sire and dam, did any of those have any sort of tooth issue?
>> ► If so, you should cull.
>> ► If not, a test breeding can be done between the individual and one of its siblings or littermates.
>> ► Inbreeding encourages negative, recessive traits to express themselves, and if offspring from the test mating show up with bad teeth, you should carefully evaluate the entire line.

At any case, any animal with bad teeth should not be sold, and defi-nitely not sold as breeding stock.

Sore Hocks

Sore hocks involve damage to the foot pad of the hind feet. It will show up initially as a rubbed spot on the foot, and will eventually progress to sores that are nearly impossible to heal. These sores can cause the rabbits extreme discomfort and can become infected.

Sore hocks can be caused by using *poor quality wire* for the pen floor. Never use fine gauge wire for floors, and definitely do not use hardware cloth or hail screen for the floor. Not only will manure not pass through the smaller square openings, the surface is rough and will constantly irritate the rabbits' feet. It is also fine gauge, and sharp.

Only use No. 0 gauge ½ × 1-inch (1.25 ×2.5 cm) mesh for the floor of a pen. It is smooth, much less sharp and will allow droppings to fall through.

If the pen is constructed of the proper wire, *conformation* is the other leading cause of sore hocks. An animal whose legs and feet do not properly line up underneath the torso will over time become sore. Evaluate any animal with hock issues to decide whether or not it should continue as breeding stock.

Resting boards and other things can be used as temporary aids, but nothing can remedy poor confirmation.

Enterotoxemia

Enterotoxemia is related to an overgrowth of *Clostridium spiroforme* in the rabbit's intestinal tract. It is most associated with feeds that are too low in fiber, when diets are changed abruptly, or after antibiotic administration. It is most prevalent in younger rabbits, three to eight weeks in age.

Prognosis is poor. Rabbits die quickly; in fact the first symptom may be a group of rabbits starting to die off. Upon necropsy, an enlarged cecum may be visible, and it may be reddish in color.

If caught early and if the rabbit will still eat, hay, rolled oats, or straw may be fed, and the rabbit can sometimes recover.

The best prevention is a good high-fiber diet, not changing diets too quickly, and avoiding stressing your rabbits out.

Enteritis

Enteritis simply means inflammation of the intestines. Most often it is related to an overgrowth of bad bacteria in the rabbit's gut. One form of enteritis, mucoid enteritis, will result in a jelly-like stool, the result of excessive mucus secretions in the rabbit's gut.

Animals with enteritis will exhibit a pot belly, and drink lots of water. In fact, you might hear a sloshing sound when you pick the rabbit up. They will be off feed, and lethargic. They will often have a dirty hind end, and will even be bloated.

Again, prognosis is poor. Necropsy will reveal an impacted cecum, with lots of jelly-like material in the intestines.

The most common cause of enteritis is a poor diet (one that is high in carbohydrates and low in fiber). A good high-fiber diet is the easiest way to prevent enteritis.

Broken Back

While a broken back is technically not an ailment but instead something that is *caused*, it is still one of the most life-threatening problems you will encounter when handling rabbits.

Rabbits have one of the lightest skeletal systems of any mammal. Compared to cats, an animal of similar size and structure, the rabbit's skeleton makes up about 7–8 percent of its weight, compared to a cat's 12–13 percent. The strong hind legs of the rabbit, combined with the light bone structure, means it is relatively easy for an improperly handled rabbit to injure itself, even breaking its back in the process.

This generally happens when restraining the animal for tattooing, for example. If the rabbit is not properly restrained so it cannot kick out with its hind legs, or if its legs are twisted when it does kick, that kick can generate enough force to break the spine, usually in the area above the pelvis.

If the injury is minor (just a bruise), the animal can recover given time and rest. In most cases, the injury is severe enough that the animal becomes paralyzed. If true paralysis is the case, prognosis is poor. The injured rabbit should be euthanized. Upon necropsy, bruising will be visible at the site of the break.

Rabbits are not typically delicate if handled properly. They can usually tolerate an accidental jump out of your arms or a tumble out of an unlatched pen door. But do keep in mind that their lighter skeleton puts them at greater risk for bone breaks, and train yourself and anyone working with them to handle them confidently and securely.

Wry Neck

Wry neck describes the peculiar and permanent head tilt that rabbits develop. In most cases, it is caused by the same bacteria that causes snuffles, *Pasturella multocida*. Other bacteria can be involved, but *Pasturella* is the most common. In this case, wry neck begins as an infection in the inner ear, which damages the ear's structure and affects the rabbit's equilibrium. It's first noticed as a slight head tilt, and when it is a full-blown case, the head can be turned completely around to the side, so the rabbit is looking straight up with one eye, and straight down with the other. This twist becomes permanent, and as treatment is ineffective, it is strongly suggested to cull the rabbit. The good news is that the meat is not affected and is still safe to eat.

Wry neck can also be caused by a stroke, and in other cases a protozoan called *Encephalitozoon cuniculi*, but the bacterial infection is the much more common cause.

In any case, while some mild cases can be treated, treatment involves long rounds of antibiotics and other drugs, which render the meat unusable. Antibiotics also have the potential to wreak havoc in the digestive system also, by indiscriminately killing healthy bacteria in the intestinal tract.

Rabbits with wry neck often manage to get around and eat and drink surprisingly well, and can grow out to the desired processing weight with little difficulty.

Coccidiosis

Coccidiosis is a disease of rabbits that is caused by a protazoan (single-celled) parasite, rather than a worm or a larvae as we typically think of parasites.

Coccidia are passed in the feces of infected animals, and require moisture and warmth to develop and be able to infect other animals.

Coccidia all belong to the genus *Eimeria*, and there are twelve different species that affect rabbits, each of which affects a different part of the intestinal system. Fortunately they all do not present the same level of threat to the host animal.

There are two main forms of coccidiosis in rabbits: an intestinal form and a hepatic (liver) form. It is very difficult to tell the two forms apart just from the symptoms; infected animals both display unthriftiness, diarrhea, poor appetite and depression. Symptoms are most pronounced in younger rabbits; adults may carry the protazoa and not have symptoms.

Four main species cause the majority of problems in the intestinal form of coccidiosis, and a single species causes the hepatic form. The hepatic form is of special concern, as the damage done to the liver, even in the moderate form, can render the rabbit unmarketable.

Treatment may or may not be successful, depending on how long the animal has been infected. Antibiotics prescribed by a veterinarian can serve to keep protazoa in check until the rabbit can develop immunity. A mild infection can result in immunity for the host animal, but immunity is species specific; immunity from one form will not give immunity from all forms of *coccidia*.

Here again, one of the best treatments is prevention. Make sure pens and all equipment is kept clean and free of feces. Don't let rabbits defecate in feeders or water bowls, keep pens clean and dry, use water bottles or automatic waterers. Don't allow animals to become overcrowded, and keep stress to a minimum. *Coccidia* will always be a greater risk to animals that are housed on the ground. Wire floors will keep fecal contamination to a minimum, and a little extra attention to keeping things clean goes a long way.

Appendix: Rabbit Breeds

Breeds on The Livestock Conservancy Conservation Priority List
Harlequin

One of the oldest breeds, references to it date back to the 1870s and 1880s. While tricolored rabbits were exhibited in Japan in the 1870s and the Harlequin was originally dubbed the "Japanese rabbit," it seems to have been first officially exhibited in France in the 1887.

The markings are critical on this rabbit, requiring alternating bands of color, and distinct separation between the colored sections.

A smaller rabbit, with adults being in the 7½ to 8-pound (3.4–3.6 kg) range, the Harlequin still has some reputation as a meat rabbit.

While it might not be suitable for intensive commercial production due to the smaller carcass, it might be very appealing to those who enjoy a good genetics challenge in getting the pattern right, and who aren't afraid to cull heavily.

Harlequin rabbits have the reputation of being relatively calm, good mothers and fairly economic eaters.

The breed is currently on TLC's Conservation Priority List in the Study category.

Belgian Hare

The Belgian Hare is the rabbit that started the rabbit craze in North America. Named for the wild hare it was created to resemble in the late 1800s, Belgian Hares were making their way from Europe to North America, and some were commanding astronomical prices.

They have a very different shape than other rabbit breeds, and are exhibited to show off that racy form. While once a very popular meat breed,

FIGURE A.1. Belgian Hare buck. Photo by Jeannette Beranger, The Livestock Conservancy. Rabbit owned by Shannon Kelly.

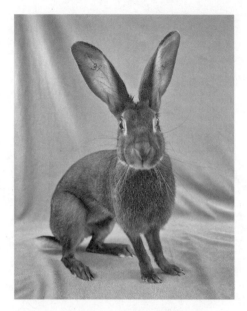

FIGURE A.2. Belgian Hare doe. Photo by Jeannette Beranger, The Livestock Conservancy. Rabbit owned by Shannon Kelly.

their survival has struggled against the development of other meatier, more commercial breeds.

The weight range for the Belgian Hare is between 6½ and 9 pounds (3–4 kg), with an ideal weight of 8 pounds (3.6 kg) for both does and bucks.

Because of their alert and active nature, Belgian Hares can be easily startled, and are probably not ideal for children or beginners. They also do not seem to be consistent in production traits.

But again, for a breeder looking for a conservation project, the Belgian might be ideal. Their smaller carcass size would be ideal for a small family, or dinner for two.

Lilac

The Lilac breed seems to have originated in several different places at nearly the same time. Ultimately all the different strains of this

FIGURE A.3. Lilac buck. Photo by Jeannette Beranger, The Livestock Conservancy. Rabbit owned by Bob and Donna Koch.

FIGURE A.4. Lilac doe. Photo by Jeannette Beranger, The Livestock Conservancy. Rabbit owned by Bob and Donna Koch.

pinkish-gray-colored rabbit merged under one name, Lilac, in 1922 and were imported to the US.

At 6 to 7 pounds (2.7–3.2 kg), it is one of the smaller meat breeds. The breed has never been hugely popular, but does have a loyal group of breeders.

They are known for being docile and good mothers, but have smaller litters.

The distinctive color of the breed makes it unique, it is truly a pinkish dove-gray color. The fur is very dense and plush, and they are truly a beautiful rabbit.

In the early years of Rare Hare, we tried Lilacs. At the time they were not successful for us, as we had not perfected our cooling system, and the animals' thick, dense fur was a problem. I do know that there are Lilac breeders all over the country in hotter climates than ours, so keep the experience of others in mind if this is a breed that interests you.

Silver

One of the oldest breeds of rabbit in the world, Silvers were already found in America at the time of the Belgian Hare

FIGURE A.5. Black Silver buck. Photo by Jeannette Beranger, The Livestock Conservancy. Rabbit owned by Jan Hall.

FIGURE A.6. Brown Silver buck. Photo by Jeannette Beranger, The Livestock Conservancy. Rabbit owned by Jan Hall.

FIGURE A.7. Fawn Silver doe. Photo by Jeannette Beranger, The Livestock Conservancy. Rabbit owned by Jan Hall.

FIGURE A.8. Blue Beveren doe. Photo by Jeannette Beranger, The Livestock Conservancy. Rabbit owned by Kim Calloway.

FIGURE A.9. White Beveren doe. Photo by Jeannette Beranger, The Livestock Conservancy. Rabbit owned by Kim Calloway.

boom of the early 1920s. The first recording of the rabbit dates back to 1631 in England, and indicates that they were well established prior to that date. The reference also indicates that their fur was especially prized and worth considerably more than other rabbit furs of the time.

Smaller in stature, Silvers are known for being docile and charismatic, good mothers, and for a small rabbit, they have good sized litters and good milking ability. The ideal weight for adult bucks and does is 6 pounds (2.7 kg). There are three color varieties: Black, Brown and Fawn.

And, as with many of the other smaller meat breeds, Silvers lost popularity when some of the larger breeds were developed.

Beveren

Beverens were developed in Belgium around the end of the 19th century, and imported into the United States in 1915. They are a large rabbit, with adults having an ideal weight of 10 pounds (4.5 kg) for bucks, and 11 pounds (5 kg) for does. They have a mandolin-shaped body, which means the loin is long and arching, resembling an overturned mandolin. There are three varieties: blue, white and black, with the blue being the original.

At one time, the blue variety was prized by furriers, having characteristics and color they desired. The white variety has distinctive blue eyes, rather than the ruby-pink eyes common to most white rabbits. Beverens have never been terribly popular, but have their own loyal following.

They are hardy, docile, and good mothers generally have large litters.

Rhinelander

The Rhinelander, as you might expect, was developed in Germany, and first exhibited in 1902. They were first brought to America in 1923, but had vanished by 1934. It is hard to say whether it was because the exacting standard of color and placement of markings was too difficult to achieve, or other larger breeds outcompeted it.

Fortunately in 1972 they were reintroduced to the US, and now enjoy a small but dedicated following.

A smaller meat breed, with ideal weight 8 to 8½ pounds (3.6–3.9 kg), Rhinelanders are alert and active, and when shown, judges are instructed not to handle them. They are allowed to pose naturally on the table, and encouraged to move. They are considered an arched breed, meaning that they should not sit flat on the table, and daylight should be visible underneath. The standard for the pattern of color is very exacting, and counts for almost two-thirds of the total for the breed standard.

While not considered ideal for commercial production due to the smaller size, the fact that the pattern is so hard to perfect means that there should be plenty of culling, and therefore lots of available fryers.

Crème d'Argent

Another French breed, the Crème d'Argent, has been bred in that country for over 150 years, possibly longer. They are a beautiful rabbit, with a cream undertone and silver ticking throughout the fur. Imported to the US before 1924, they were crossed with other meat-type breeds to further develop the breed's meat qualities. Today the breed is extinct in all countries except the United States and England.

A large breed, Crème d'Argent rabbits are docile and usually good mothers. Ideal weights for bucks and does are 9 and 10 pounds (4 and 4.5 kg), respectively.

FIGURE A.10. Rhinelander doe. Photo by Jeannette Beranger, The Livestock Conservancy. Rabbit owned by Mikayla Herber.

FIGURE A.11. Crème d'Argent doe. Photo by Jeannette Beranger, The Livestock Conservancy. Rabbit owned by Mark Cox.

FIGURE A.12. American Blue doe. Photo by The Rare Hare Barn. Rabbit owned by The Rare Hare Barn.

FIGURE A.13. American White doe. Photo by The Rare Hare Barn. Rabbit owned by The Rare Hare Barn.

American Blue and White

The American Blue and White is one of the four breeds on the Livestock Conservancy CPL that are unique to North America. The Blue was developed here and recognized as a breed in 1918. Only the New Zealand Red predates the American Blue in development, and not by much.

A number of other blue varieties of rabbit existed at the time, and blue seemed to be a popular color for fanciers.

The breed standard was written for a single rabbit owned by Lewis Salisbury of Pasadena, California, and for many years his blues were considered the finest in the country.

The American Blue was considered to have the best blue color of all the blue varieties, and furriers were paying an incredible $2 per pelt in 1920, and a good breeding doe with a pedigree could fetch as much as $25.

The American Blue has a mandolin shape, and its loin is longer than that of non-mandolin breeds.

The white variety followed in 1925.

At one time the American Blue and White was the rarest of the uniquely American breeds, but it seems to have gained ground over recent years. The blue variety remains the more popular, because of the beautiful color, and the white variety is always in need of good stewards.

They are docile and even-tempered, good mothers and have good size litters. They are large: bucks weighing from 9 to 11 pounds (4–5 kg), and does from 10 to 12 pounds (4.5–5.4 kg).

American Chinchilla

The American Chinchilla is another breed unique to North America.

The Standard Chinchilla originated in France, but soon found its way to the US. In 1919, Edward Stahl purchased all the stock available at the New York State Fair. Stahl and several other breeders set about using the largest of the Standard Chinchilla to create the American

Heavyweight Chinchilla, which ultimately became the American Chinchilla.

Selected to resemble the Chinchilla of South America, this rabbit's fur caught the world by storm. The fur at first appears to be salt and pepper colored, but close inspection reveals four distinct color bands.

The breed had a huge impact on the rabbit world, and has given rise to numerous other breeds. Between November of 1928 and November 1929, 17,328 Chinchilla were registered with ARBA, a record which remains unbroken.

The American Chinchilla is an outstanding meat rabbit, producing uniform litters that reach market weight quickly. They are good mothers, and while very active, vigorous rabbits, tolerate handling easily.

They are large rabbits: bucks in the 9 to 11 pound (4–5 kg) range and does in the 10 to 12 pound (4.5–5.4 kg) range.

FIGURE A.14. American Chinchilla doe. Photo by The Rare Hare Barn. Rabbit owned by The Rare Hare Barn.

Blanc de Hotot

The name literally means "White from Hotot," Hotot being a small region in Northwestern France where this breed was developed.

Hotot have the distinction of being one of the few breeds of rabbit developed by a female breeder. Beginning in 1902 with a breed that was spotted, and crossing with white breeds until the spots were dramatically reduced, it took hundreds of breedings, but Madame Eugenie Bernhard finally realized her goal of a white rabbit with brown eyes. Almost. The white rings around the eyes proved nearly impossible to remove, and the Hotot retains this distinctive feature today.

The Hotot was finally recognized by the French as a breed in 1922. The first Hotots were imported into the United States in 1978. Even though they are still considered rare, they have always been in demand because of their striking appearance.

In our experience they have not always been the best production animals, tending to be on the nervous side, but they have fine carcass quality.

Does range from 9–11 pounds (4–5 kg), and bucks from 8–10 pounds (3.6–4.5 kg).

FIGURE A.15. Blanc de Hotot doe. Photo by Jeannette Beranger, The Livestock Conservancy. Rabbit owned by Barb Semb.

FIGURE A.16. Silver Fox rabbit. Photo by The Rare Hare Barn. Rabbit owned by The Rare Hare Barn.

Silver Fox

The third of the uniquely American breeds is the Silver Fox. Recognized in 1925, the breed originally came in two varieties, black and blue. It is solid colored, with white hairs scattered throughout the pelt.

The breed very nearly became extinct, but a group of breeders formed the National Silver Fox Rabbit Club in 1971. Unfortunately the blue variety was dropped due to lack of rabbits shown at ARBA shows within a five-year period, and has yet to be reinstated.

The breed has beautiful, lustrous fur, and the fur is unique in that when stroked from tail to head, it should remain upright until stroked back the other way.

They are also an excellent meat rabbit, generally good mothers and have nice litters that grow quickly. We have not found them to dress at 65 percent regularly, but their dress out percentage is certainly one of the best.

They are another large breed, the ideal weight for bucks is 9½ pounds, and does 10½ pounds (4.3–4.8 kg).

Breeds Not on the Conservancy Priority List
Palomino

The Palomino is an American creation, developed in the 1950s from a mixture of sources. They are a good, solid meat and fur breed, good mothers and with a good growth rate. They are one of the larger breeds, with does ideal weight at 10 pounds (4.5 kg), and bucks at 9 pounds (4 kg).

Rex

The Rex is unique in the rabbit world by virtue of their fur. Resulting from a mutation which produced coats without the longer guard hairs, Rex fur has a plush, almost velvety feel. The Rex fur took some time to be developed and perfected, but now there are many color varieties and patterns to choose from.

Fur is the major defining characteristic of this breed, but they are also good meat rabbits as well. The ideal weight for does is 9 pounds (4 kg), bucks 8 pounds (3.6 kg).

New Zealand

Despite its name, the New Zealand rabbit is an American creation. The breed was developed at the same time in the late 1900s in several locations in the US, and as the Belgian Hare boom waned, the New Zealand quickly assumed the role of the premiere meat rabbit in its day.

The red was the first variety created and was quite popular, given it's beautiful rich red coloring coupled with outstanding meat qualities, but the white soon supplanted it, as processors began preferring the white rabbits over the colored varieties of many breeds.

New Zealands are a large rabbit, good mothers and have good litter size. Their ideal weights are 10 pounds (4.5 kg) for bucks and 11 pounds (5 kg) for does.

Satin

The Satin breed, like the Rex, was developed around a single gene mutation, found in a single rabbitry in 1934. The unique sheen of the fur of these rabbits caught on quickly, and fanciers enjoyed playing with different colors.

The hair shaft of the Satin is more transparent, and of finer diameter, than on other rabbits, allowing the pigment to appear more clearly. Satins have more vibrant color than normal furred breeds, and a higher sheen to the coat.

While known for the fur, Satins are an excellent meat type as well, ranging from 9–11 pounds (4–5 kg) adult weight.

Californian

The Californian was developed over several years with the goal of being an outstanding meat and fur breed. The Californian was given a working standard

FIGURE A.17. Champagne d'Argent buck. Photo by Jeannette Beranger, The Livestock Conservancy. Rabbit owned by Lisa Schmidt Daughtery.

by ARBA in 1939. It is recognized worldwide as an outstanding meat rabbit. Ideal weight for bucks is 9 pounds (4 kg), for does 9½ pounds (4.3 kg).

Cinnamon

The Cinnamon is a relative newcomer to the list of rabbit breeds, but it has a loyal following who appreciate its meat qualities. The breed got its start as a 4-H project. As the name suggests, it is a cinnamon, or rust color. Does average 10 pounds (4.5 kg), and bucks 9½ pounds (4.3 kg)

Champagne d'Argent

The Champagne d'Argent has the distinction of being considered the oldest recognized breed of rabbit. References to them date back to 1631, and it had likely been bred by monks in the Champagne valley of France for hundreds of years prior.

Champagnes are a beautiful, uniform silver color all over, and back in the day, peasants could make a sizeable amount from the sale of the furs. No special mention is given to the rabbit, as they were considered the common stock of the region, but fur buyers from all over came to purchase large quantities of the pelts.

They are an excellent meat breed, and good mothers. Does range in weight from 9½ to 12 pounds (4.3–5.4 kg), bucks from 9 to 11 (4–5 kg).

Endnotes

Chapter 1

1. Nieves Lopez-Martinez. "The Lagomorph Fossil Record and the Origin of the European Rabbit" in P. C. Alves et al., eds. *Lagomorph Biology*. Springer, 2008. [online]. [cited June 28, 2018]. doi.org/10.1007/978-3-540-72446-9_3.

2. Francois Lebas. "Origine du lapin et domestication" in *Historique de l'élevage du lapin*. [online]. [cited July 1, 2018]. cuniculture.info/Docs/Elevage/Histori-01.htm#2.

3. Bob D. Whitman. *Domestic Rabbits & Their Histories: Breeds of the World*. Leathers, 2004, p. 344.

4. Lebas, "Passage de l'animal sauvage élevé en enclos au lapin domestique" in *Historique de l'élevage du lapin*.

5. Ibid.

6. Whitman, p. 346.

7. Lebas, "Passage de l'animal sauvage élevé en enclos au lapin domestique."

8. Ibid.

9. Ibid.

10. Whitman, p. 70.

11. Whitman, pp. 368–376.

12. Whitman, pp. 45, 129, and 314.

13. J. C. Fehr, ed. *Hares and Rabbits*, Vol. 4 (August 1919).

14. James I. McNitt et al. *Rabbit Production*, 8th ed. Interstate Publishers, 2000, p. 29.

Chapter 2

1. USDA Food Safety and Inspection Service. *Rabbit from Farm to Table*. [online]. [cited July 13, 2018]. fsis.usda.gov/wps/portal/fsis/topics/food-safety-education/get-answers/food-safety-fact-sheets/meat-preparation/rabbit-from-farm-to-table/rabbit-farm-table.

2. Gleaned from our experience over the years, and conversations with our friends at The Livestock Conservancy.

Chapter 3

1. McNitt et al., *Rabbit Production*, p. 5.
2. Ibid, p. 6; and National Museum Australia. *Rabbits in Australia*. [online]. [cited July 9, 2018]. nma.gov.au/online_features/rabbits_in_australia.
3. McNitt et al., p. 455.
4. McNitt et al., p. 279; Francois Lebas et al. *The Rabbit: Husbandry, Health and Production*. UN Food and Agriculture Organization, 1997, p. 22. [online]. [cited July 1, 2018]. fao.org/docrep/t1690e/t1690e00.htm.

Chapter 4

1. The Livestock Conservancy website. [online]. [cited May 17, 2018]. livestockconservancy.org.
2. American Rabbit Breeders Association website. [online]. [cited May 17, 2018]. arba.net.
3. The Livestock Conservancy. "Conservation Priority List." [online]. [cited June 26, 2018]. livestockconservancy.org/index.php/heritage/internal /conservation-priority-list.
4. D. Phillip Sponenberg, Jeanette Beranger, and Alison Martin. *An Introduction to Heritage Breeds*. Storey, 2014, p. 156.
   A good general reference for all the breeding strategies is: D. Phillip Sponenberg, Jeanette Beranger, and Alison Martin. *Managing Breeds for a Secure Future: Strategies for Breeders and Breed Associations*, 2nd ed. 5M Publishing, 2017.

Chapter 5

1. See Chapter 6, "Sanitation."
2. American Rabbit Breeders Association. "ARBA Recognized Breeds." [online]. [cited July 3, 2018]. arba.net/breeds.htm.

Chapter 6

1. Search "solar power" and "wind power" at New Society's website: newsociety.com.

Chapter 7

1. An intestinal disease caused by parasites. See Chapter 15 for more information.

Chapter 8

1. Lebas et al., *The Rabbit: Husbandry, Health and Production*, pp. 19, 22.

2. Susan Lumpkin 'and John Seidensticker. *Rabbits: The Animal Answer Guide*. Johns Hopkins, 2011, p.36.

3. Eric and Callene Rapp. Sustainable Agriculture Research and Education (SARE) North Central grant, Project Number FCN06-604 (March 23, 2007 through December 31, 2008).

Chapter 9

1. See American Rabbit Breeders Association. "ARBA Recognized Breeds." [online]. [cited July 3, 2018]. arba.net/breeds.htm; an older edition is Glen C. Carr. *Standard of Perfection*. American Rabbit Breeders Association, 2006.

Chapter 10

1. McNitt et al., *Rabbit Production*, p. 79.

Chapter 11

1. Lebas et al., *The Rabbit: Husbandry, Health and Production*, p. 50.

2. McNitt et al., *Rabbit Production*, p. 155.

Chapter 12

1. Evans Software Services. "Rabbit Breeders Products and Services." [online]. [cited May 30, 2018]. evsoft.us.

Chapter 13

1. Rabbit Wringer—Rabbit and Poultry Harvesting Equipment. [online]. [cited July 2, 2018]. rabbitwringer.com.

2. Adam Danforth. *Butchering Poultry, Rabbit, Lamb, Goat, and Pork: The Comprehensive Photogenic Guide to Humane Slaughtering and Butchering*. Storey, 2014, p. 20.

3. Ibid., p. 33.

4. Ibid., p. 25.

5. Ibid, p. 26.

6. Andrea Rock. "How Safe Is Your Ground Beef?" *Consumer Reports*, December 21, 2015. [online]. [cited July 2, 2018]. consumerreports.org /cro/food/how-safe-is-your-ground-beef.

Chapter 14

1. USDA Agricultural Research Service, National Nutrient Database for Standard Reference Legacy Release. "Basic Report: 17178, Game meat,

rabbit, domesticated, composite of cuts, cooked, roasted." [online]. [cited July 10, 2018]. ndb.nal.usda.gov/ndb/foods/show/17178; USDA Agricultural Research Service, National Nutrient Database for Standard Reference Legacy Release. "Basic Report: 13922, Beef, tenderloin, roast, separable lean and fat, trimmed to ⅛" fat, choice, cooked, roasted." [online]. [cited July 10, 2018]. ndb.nal.usda.gov/ndb/foods/show/302961.

2. USDA Food Safety and Inspection Service, *Rabbit from Farm to Table*.

Chapter 15

1. For more more information on conditions described in this chapter see: Lebas et al., *The Rabbit: Husbandry, Health and Production*; McNitt et al., *Rabbit Production*; Lumpkin and Seidensticker, *Rabbits: The Animal Answer Guide*; George S. Templeton, *Domestic Rabbit Production*. Interstate, 1968; Chris Hayhow, *Care of the Domestic Rabbit*. Leathers, 2003.

2. Lebas et al., *The Rabbit: Husbandry, Health and Production*, p. 112.

Index

About the Authors

Since 2005, Eric and Callene Rapp have owned and operated the award-winning Rare Hare Barn, the largest heritage-breed meat-rabbit enterprise in the United States, supplying heritage breed fryers to local restaurants. They have over 50 years of combined experience handling nearly every species of domestic livestock, and are active members of The Livestock Conservancy. Callene is also a regular contributor to *Grit* magazine. They live and farm in Leon, Kansas. rareharebarn.com

ABOUT NEW SOCIETY PUBLISHERS

New Society Publishers is an activist, solutions-oriented publisher focused on publishing books for a world of change. Our books offer tips, tools, and insights from leading experts in sustainable building, homesteading, climate change, environment, conscientious commerce, renewable energy, and more—positive solutions for troubled times.

We're proud to hold to the highest environmental and social standards of any publisher in North America. This is why some of our books might cost a little more. We think it's worth it!

- We print all our books in North America, never overseas

- All our books are printed on **100% post-consumer recycled paper**, processed chlorine-free, with low-VOC vegetable-based inks (since 2002)

- Our corporate structure is an innovative employee shareholder agreement, so we're one-third employee-owned (since 2015)

- We're carbon-neutral (since 2006)

- We're certified as a B Corporation (since 2016)

At New Society Publishers, we care deeply about *what* we publish—but also about *how* we do business.

Download our catalog at https://newsociety.com/Our-Catalog or for a printed copy please email info@newsocietypub.com or call 1-800-567-6772 ext 111.

New Society Publishers
ENVIRONMENTAL BENEFITS STATEMENT

For every 5,000 books printed, New Society saves the following resources:[1]

25	Trees
2,291	Pounds of Solid Waste
2,521	Gallons of Water
3,288	Kilowatt Hours of Electricity
4,164	Pounds of Greenhouse Gases
18	Pounds of HAPs, VOCs, and AOX Combined
6	Cubic Yards of Landfill Space

[1] Environmental benefits are calculated based on research done by the Environmental Defense Fund and other members of the Paper Task Force who study the environmental impacts of the paper industry.